D1348922

Nitrogen Fixation and CO_2 Metabolism

Nitrogen Fixation and CO_2 Metabolism

A Steenbock Symposium in Honor of Professor Robert H. Burris

Proceedings of the Fourteenth Steenbock Symposium held 17–22 June 1984 at the University of Wisconsin–Madison, Madison, Wisconsin, U.S.A.

Editors

Paul W. Ludden, Ph.D.
Assistant Professor of Biochemistry, Department of Biochemistry, College of Agricultural and Life Sciences, University of Wisconsin–Madison, Madison, Wisconsin, U.S.A.

and

John E. Burris, Ph.D.
Associate Professor of Biology, Department of Biology, College of Science, The Pennsylvania State University, University Park, Pennsylvania, U.S.A.

Elsevier
New York • Amsterdam • Oxford

Elsevier Science Publishing Co., Inc.
52 Vanderbilt Avenue, New York, New York 10017

Distributors outside the United States and Canada:

Elsevier Science Publishers B.V.
P.O. Box 211, 1000 AE, Amsterdam, The Netherlands

Library of Congress Cataloging in Publication Data

Steenbock Symposium (14th: 1984: University of Wisconsin—Madison)

Nitrogen fixation and CO_2 metabolism.
"A symposium in honor of professor Robert H. Burris."

Bibliography: p.
Includes indexes.
1. Nitrogen—Fixation—Congresses. 2. Carbon dioxide—Metabolism—Congresses. 3. Micro-organisms, Nitrogen-fixing—Congresses. 4. Plants—Metabolism—Congresses. I. Ludden, Paul W. II. Burris, John E. III. Burris, Robert H. (Robert Harza), 1914- . IV. Title.
QR89.7.S77 1984 589.9′504133 84-24735
ISBN 0-444-00953-1

Manufactured in the United States of America

Contents

PREFACE

Photosynthesis and nitrogen fixation are key metabolic processes which lead to the production of reduced carbon and nitrogen compounds. These compounds are essential for the maintenance and continuation of life on earth. In this volume many of the latest advances in the study of nitrogen fixation and photosynthetic carbon dioxide fixation are presented.

These two topics were chosen as the subjects for the 14th Annual Harry Steenbock Symposium for their importance and because they represent the two main research areas of Professor Burris, one of the most productive and innovative researchers in plant biochemistry. In his unswerving dedication to unraveling the mysteries surrounding these two basic processes, Professor Burris has elucidated numerous aspects of nitrogen fixation and photosynthesis. He has also trained an extraordinary number of undergraduates, graduates, and postdoctoral fellows to be better and more careful scientists.

We hope you will enjoy reading the chapters written by the participants in this symposium. These papers include topics ranging from molecular genetics through crop breeding and enzymology through crop productivity. A variety of organisms are examined as the authors present their research on the diverse array of organisms engaged in nitrogen fixation and photosynthetic carbon dioxide metabolism.

This symposium series honors the memory of the distinguished University of Wisconsin biochemist Harry Steenbock. The 14th Annual Harry Steenbock Symposium was supported by funds provided by Dr. Evelyn Steenbock (Mrs. Harry) to the Department of Biochemistry, College of Agricultural and Life Sciences of the University of Wisconsin-Madison, the United States Department of Energy, E. I. duPont de Nemours, Monsanto, Biotechnology General, Exxon Research and Engineering, Cetus-Madison, and Nitragin. In addition to our thanks to the above for financial support, we would particularly like to thank the speakers, session chairmen (Harry Winter, Vinod Shah, Jerald Ensign, Bruce Selman, Lawrence Schrader, Marion O'Leary), and Karen Davis whose tireless efforts in organizing the symposium and editing this volume were crucial to their success.

Finally, our greatest debt of gratitude is owed to Professor Robert H. Burris who, for almost 50 years, has upheld the highest standards of scientific investigation. His dedication, skill, and integrity have served as examples for all of us to attempt to emulate.

This volume and the symposium are our humble means of thanking Robert H. Burris for all his contributions.

John E. Burris
Paul W. Ludden

Symposium photos by Raymond E. Smith

Students and Associates of R. H. Burris Present at the Conference:

1. M. J. Dilworth
2. Leonard E. Mortenson
3. Linda Jo Rasmussen
4. Robert E. Stutz
5. Laura S. Privalle
6. Haian Fu
7. Paul W. Ludden
8. Diana Pitterle
9. David W. Emerich
10. Richard B. Peterson
11. Ayala Hochman
12. John DuBois
13. Israel Zelitch
14. N. E. Tolbert
15. R. H. Burris
16. Lee C. Olson
19. Thomas K. Soulen
21. Y. I. Shethna
22. Vladimir Zbinovsky
23. Wilbur H. Campbell
24. Jeffrey P. Houchins
25. Ted Stern
26. Cheryl Burpee
27. Joseph H. Guth
29. Anton Hartmann
30. Jane Hasselkus
31. Robert V. Hageman
32. Yaacov Okon
33. Daniel J. Arp
34. John D. Tjepkema
35. Jihong Liang
36. Harry C. Winter
37.
38. David R. Benson
39. Robert V. Klucas
40. Marion O'Leary
41. W. D. P. Stewart
42. Seung-Dal Song
43. Barry Osmond
44. J. W. Newton
45. Frank Simpson

PARTICIPANTS

Marilyn Abbott
Department of Biochemistry
University of Wisconsin-Madison
Madison, WI 53706, USA

Edward Appelbaum
Agrigenetics Corporation
Advanced Research Division
5649 E. Buckeye Rd.
Madison, WI 53714, USA

Cyril A. Appleby
Division of Plant Industry
CSIRO
GPO Box 1600
Canberra, AUSTRALIA 2601

Jay Arnone
Yale School of Forestry
and Environmental Studies
370 Prospect Street
New Haven, CT 06511, USA

Daniel J. Arp
Department of Biochemistry
University of California
Riverside, CA 92521, USA

Sandra G. Auger
Allied Corporation
P.O. Box 6
Solvay, NY 13209, USA

Peter Avtges
CLSC 355
University of Chicago
520 E. 48th St.
Chicago, IL 60637, USA

Dwight Baker
Battelle-C.F. Kettering
Research Laboratory
150 E. South College St.
Yellow Springs, OH 45387, USA

Judith Bale
Board on Science and Technology
for International Development
National Academy of Sciences
2101 Constitution Ave., NW
Washington, D.C. 20418, USA

Susan R. Barnum
Department of Genetics
Iowa State University
8 Curtiss
Ames, IA 50011, USA

Terese Barta
Department of Plant Pathology
University of Wisconsin-Madison
Madison, WI 53706, USA

Chris Baysdorfer
USDA/ARS
Light and Plant Growth Lab
BARC-W
Beltsville, MD 20705, USA

Dwight U. Beebe
Botany Department
University of Wisconsin-Madison
Madison, WI 53706, USA

David R. Benson
Microbiology Section
University of Connecticut
Storrs, CT 06268, USA

David L. Berryhill
Department of Bacteriology
North Dakota State University
Fargo, ND 58105, USA

Paul E. Bishop
Department of Microbiology
North Carolina State University
P.O. Box 7615
Raleigh, NC 27695, USA

Clanton C. Black, Jr.
Department of Biochemistry
University of Georgia
Athens, GA 30602, USA

Fredrick A. Bliss
Department of Horticulture
University of Wisconsin-Madison
Madison, WI 53706, USA

Duane Bonam
Department of Biochemistry
University of Wisconsin-Madison
Madison, WI 53706, USA

Roberto Borghese
Biochemistry Department
322-A Chemistry Building
University of Missouri-Columbia
Columbia, MO 65211, USA

Winston J. Brill
Department of Bacteriology
and Center for Studies of
Nitrogen Fixation
University of Wisconsin-Madison
Madison, WI 53706, USA

Susan J. Brown
Department of Biochemistry
Willard Hall
Kansas State University
Manhattan, KS 66502, USA

John E. Burris
Department of Biology
202 Buckhout Laboratory
The Pennsylvania State University
University Park, PA 16802, USA

Robert H. Burris
Department of Biochemistry
University of Wisconsin-Madison
Madison, WI 53706, USA

Joe C. Burton
Niftal
University of Hawaii
4930 N. Elkhart Ave.
Milwaukee, WI 53217, USA

Brian Buttery
Agriculture Canada
Research Station
Harrow, Ontario
CANADA NOR 1GO

Pamela J. Camp
Department of Biochemistry
322-A Chemistry Building
University of Missouri
Columbia, MO 65211, USA

Wilbur H. Campbell
Department of Chemistry
State University of New York
College of Environmental Science
and Forestry
Syracuse, NY 13210, USA

Russell W. Carlson
Department of Chemistry
Eastern Illinois University
Charleston, IL 61920, USA

Todd A. Carlson
MSU-DOE Plant Research Laboratory
Michigan State University
East Lansing, MI 48824, USA

Sergio Casella
Istituto di Microbiologia Agraria
Universita di Pisa
Centro di Studio per la
Microbiologia del Suolo, CNR
Pisa, ITALY

Joseph Cava
Department of Biology
Marquette University
Milwaukee, WI 53233, USA

Christina Chang
Department of Biochemistry
Kansas State University
Manhattan, KS 66502, USA

Nicole Chartrain
Agrigenetics Corporation
Advanced Research Division
5649 E. Buckeye Rd.
Madison, WI 53706, USA

John Chisnell
Department of Microbiology
North Carolina State University
Box 7615
Raleigh, NC 27695-7615, USA

Winnie Chow
Genetics Department
Iowa State University
Ames, IA 50011, USA

A. Lawrence Christy
Agricultural Research Division
Monsanto Company
St. Louis, MO 63167, USA

Janet L. Clawitter
Department of Biochemistry
and Interdisciplinary Plant
Biochemistry/Physiology Group
University of Missouri
Columbia, MO 65211, USA

David A. Dalton
Department of Botany and
Laboratory for Nitrogen
Fixation Research
Oregon State University
Corvallis, OR 97331, USA

Lawrence C. Davis
Department of Biochemistry
Kansas State University
Manhattan, KS 66506, USA

Frank B. Dazzo
Department of Microbiology
and Public Health
Michigan State University
East Lansing, MI 48824, USA

Thomas Lloyd Deits
Department of Biochemistry
University of Minnesota
4-235 Millard Hall
435 Delaware St., S.E.
Minneapolis, MN 55455-0326, USA

Eliane Deleens
Laboratoire du Phytotron
91190 Gif-sur-Yvette
FRANCE

Richard E. Dickson
Forestry Sciences Laboratory
P.O. Box 898
Rhinelander, WI 54501, USA

Ronald Diebold
Wehr Life Science Building
Marquette University
Milwaukee, WI 53233, USA

M. J. Dilworth
Nitrogen Fixation Research Group
School of Environmental and Life Sciences
Murdoch University
Murdoch, W. AUSTRALIA 6150

Charles M. Doyle
Department of Biochemistry
University of California
Riverside, CA 92521, USA

John D. DuBois
Department of Biochemistry
University of Wisconsin-Madison
Madison, WI 53706, USA

Stanley H. Duke
Department of Agronomy
University of Wisconsin-Madison
Madison, WI 53706, USA

Nigel S. Dunn-Coleman
Central Research & Development Department
E. I. duPont de Nemours & Company
Experimental Station
Wilmington, DE 19898, USA

Allan R. J. Eaglesham
Boyce Thompson Institute for Plant Research at
Cornell University
Ithaca, NY 14853, USA

Stephen W. Ela
Venture Science Consulting
1240 Sherman Avenue
Madison, WI 53703, USA

David W. Emerich
Department of Biochemistry
322A New Chemistry Building
University of Missouri
Columbia, MO 65211, USA

Nila J. Emerich
Department of Biochemistry
322A New Chemistry Building
University of Missouri
Columbia, MO 65211, USA

Kim Stutzman Engwall
Department of Genetics
Iowa State University
Ames, IA 50011, USA

Jerald C. Ensign
Department of Bacteriology
University of Wisconsin-Madison
Madison, WI 53706, USA

B. Kipp Erickson
Department of Agricultural Biochemistry
University of Nebraska
Lincoln, NE 68583-0718, USA

Harold J. Evans
Laboratory for Nitrogen Fixation Research
Oregon State University
Corvallis, OR 97331

Wayne P. Fitzmaurice
Department of Bacteriology
University of Wisconsin-Madison
Madison, WI 53706, USA

George Flentke
Department of Biochemistry and Institute for Enzyme Research
University of Wisconsin-Madison
Madison, WI 53706, USA

Enrique Flores
MSU-DOE
Plant Research Laboratory
Michigan State University
East Lansing, MI 48824

Luiji Frusciante
Department of Horticulture
University of Wisconsin-Madison
Madison, WI 53706, USA

Haian Fu
Department of Biochemistry
University of Wisconsin-Madison
Madison, WI 53706, USA

Martin Fuhrmann
Mikrobiologisches Institut
Eidgenossische Technische Hochschule
ETH-Zentrum
CH-8092 Zurich, SWITZERLAND

Essam M. Gewaily
Agriculture Botany Department
College of Agriculture
Zagazig University
Zagazig, EGYPT

Champa Sengupta Gopalan
Agrigenetics Corporation
Advanced Research Division
5649 E. Buckeye Rd.
Madison, WI 53714, USA

Joyce K. Gordon
Dynacomp Research Corporation
802-111 W. Hastings
Vancouver, British Columbia
CANADA V6E 2J3

Joseph H. Guth
66 Rivington St., #18
New York, NY 10002, USA

Huub Haaker
Department of Biochemistry
Agricultural University
De Dreijen 11
6703 BC Wageningen
THE NETHERLANDS

Robert V. Hageman
Synergen
1885 3rd Street
Boulder, CO 80301, USA

Larry J. Halverson
Department of Microbiology
University of Tennessee
Knoxville, TN 37996, USA

Arnold Hampel
Biology Department
Northern Illinois University
Dekalb, IL 60115, USA

R. W. F. Hardy
Central Research & Development Department
E. I. du Pont de Nemours & Company
Experimental Station E328/422
Wilmington, DE 19898, USA

Alan R. Harker
Laboratory for Nitrogen Fixation Research
Oregon State University
Corvallis, OR 97331, USA

Anton Hartmann
Department of Biochemistry
University of Wisconsin-Madison
Madison, WI 53706, USA

Robert Haselkorn
University of Chicago
920 E. 58th Street
Chicago, IL 60637, USA

Rex Hayes
Department of Biochemistry
322A New Chemistry Building
University of Missouri
Columbia, MO 65211, USA

John P. Helgeson
Department of Plant Pathology
University of Wisconsin-Madison
Madison, WI 53706, USA

Cynthia A. Henson
Department of Agronomy
University of Wisconsin-Madison
Madison, WI 53706, USA

Antonia Herrero
MSU-DOE
Plant Research Laboratory
Michigan State University
East Lansing, MI 48824, USA

Ayala Hochman
Department of Biochemistry
Tel-Aviv University
Ramat-Aviv, ISRAEL

Adrian L. M. Hodgson
Department of Microbiology
Walters Life Sciences Building
University of Tennessee
Knoxville, TN 37996, USA

Alan Hooper
Department of Genetics and
Cell Biology
University of Minnesota
St. Paul, MN 55108, USA

Timothy Hoover
Department of Biochemistry
University of Wisconsin-Madison
Madison, WI 53706, USA

Jeffrey P. Houchins
Department of Botany
University of Minnesota
220 Biosciences Center
St. Paul, MN 55108, USA

James Bryant Howard
Department of Biochemistry
4-225 Millard Hall
University of Minnesota
435 Delaware St., S.E.
Minneapolis, MN 55455-0326, USA

Steven C. Huber
USDA/ARS and
Department of Crop Science
North Carolina State University
Raleigh, NC 27695-7631, USA

Kerstin Huss-Danell
Department of Plant Physiology
University of Umea
S-90187 Umea, SWEDEN

Juan R. Imperial
Department of Bacteriology
University of Wisconsin-Madison
Madison, WI 53706, USA

John Imsande
Department of Genetics
Iowa State University
Ames, IA 50011, USA

Robert J. Ireland
Biology Department
Carleton University
Ottawa, Ontario
CANADA K1S 5B6

Yehia Zaki Ishac
Unit of Bio-Fertilizers
Faculty of Agriculture
Ain-Shams University
Cairo, EGYPT

Marty Jacobson
Department of Microbiology
North Carolina State University
Box 7615
Raleigh, NC 27695-7615

Mahendra K. Jain
Department of Bacteriology
University of Wisconsin-Madison
Madison, WI 53706, USA

Rolf D. Joerger
Department of Microbiology
North Carolina State University
Box 7615
Raleigh, NC 27695-7615, USA

Eric Johansen
Agrigenetics Corporation
Advanced Research Division
5649 E. Buckeye Rd.
Madison, WI 53714, USA

Benjamin L. Jones
Lipid Metabolism Lab
V.A. Hospital
2500 Overlook Terrace
Madison, WI 53705, USA

Ken W. Joy
Department of Biology
Carleton University
Ottawa, Ontario
CANADA K1S 5B6

Michael Kahn
Department of Bacteriology
Washington State University
Pullman, WA 99164, USA

Barbara J. Kamicker
Department of Bacteriology
University of Wisconsin-Madison
Madison, WI 53706, USA

Yasuko Kaneko
Department of Botany
University of Wisconsin-Madison
Madison, WI 53706, USA

Roy Kanemoto
Department of Biochemistry
University of Wisconsin-Madison
Madison, WI 53706, USA

T. Kaneshiro
USDA/ARS
Northern Regional Research
Center
Peoria, IL 61604, USA

Lawrence A. Kapustka
Botany Department
Miami University
Oxford, OH 45056, USA

Dale B. Karr
Department of Biochemistry
322A Chemistry Building
University of Missouri
Columbia, MO 65211, USA

Kenneth Keegstra
Department of Botany
University of Wisconsin-Madison
Madison, WI 53706, USA

Donald L. Keister
C.F. Kettering Research Laboratory
150 E. South College St.
Yellow Springs, OH 45387, USA

Leszek A. Kleczkowski
Department of Biochemistry
322A Chemistry Building
University of Missouri
Columbia, MO 65211, USA

Robert V. Klucas
Department of Agricultural Biochemistry
University of Nebraska-Lincoln
Lincoln, NE 68583, USA

Adam Kondorosi
Institute of Genetics
Biological Resources Center
Hungarian Academy of Sciences
P.O. Box 521
H-6702 Szeged, HUNGARY

Eva Kondorosi
Institute of Genetics
Biological Resources Center
Hungarian Academy of Sciences
P.O. Box 521
H-6701 Szeged, HUNGARY

Robert G. Kranz
CLSC Rm. 355
University of Chicago
920 E. 48th St.
Chicago, IL 60637, USA

Bruce Kulpaca
Department of Biology
Marquette University
Milwaukee, WI 53233, USA

David H. Lambert
Microbiology Program
The Pennsylvania State University
University Park, PA 16802, USA

Grant R. Lambert
Laboratory for Nitrogen Fixation Research
Oregon State University
Corvallis, OR 97331, USA

Miguel Lara
Centro de Investigacion sobre Fijacion de Nitrogeno
Universidad Nacional Autonoma de Mexico
Apdo. Postal 565-A
Cuernavaca, Mor. MEXICO

T. A. LaRue
Boyce Thompson Institute
Tower Road
Ithaca, NY 14853, USA

Keuk-Ki Lee
Department of Agricultural Biochemistry
University of Nebraska
Lincoln, NE 58583, USA

Craig Lending
Department of Botany
University of Wisconsin-Madison
Madison, WI 53706, USA
(608-262-3667)

Sally Leong
Department of Plant Pathology
University of Wisconsin-Madison
Madison, WI 53706, USA

Walter T. Leps
Alberta Research council
11315 87th Ave.
Edmonton, Alberta
CANADA T6K OY8

Jihong Liang
Department of Biochemistry
University of Wisconsin-Madison
Madison, WI 53706, USA

R.-T. Liang
Department of Biochemistry
322A Chemistry Building
University of Missouri
Columbia, MO 65211, USA

Mary E. Lidstrom
Department of Microbiology SC-42
University of Washington
Seattle, WA 98195, USA

James M. Ligon
Allied Corporation
P.O. Box 6
Solvay, NY 13209, USA

Sharon R. Long
Department of Biological Sciences
Stanford University
Stanford, CA 94305, USA

Mary F. Lopez
Cabot Foundation
Harvard University
Petersham, MA 01366, USA

Jean Love
Biochemistry Department
Rm. 322 New Chemistry Building
University of Missouri
Columbia, MO 65211, USA

Bob Lowery
Department of Biochemistry
University of Wisconsin-Madison
Madison, WI 53706, USA

Paul W. Ludden
Department of Biochemistry
University of Wisconsin-Madison
Madison, WI 53706, USA

Jean Lukens
The Biological Laboratories
Harvard University
Cambridge, MA 02138, USA

Michael T. Madigan
Department of Microbiology
Southern Illinois University
Carbondale, IL 62901, USA

R. J. Maier
Department of Biology
Mudd Hall
Johns Hopkins University
34th & Charles
Baltimore, MD 21218, USA

Iris F. Martin
USDA Competitive Research Grants Office
Suite 112 West Auditors Bldg.
15th St. & Independence Ave., S.W.
Washington, D.C. 20251, USA

Hisashi Matsushima
Department of Botany
University of Wisconsin-Madison
Madison, WI 53706, USA

Tom McLoughlin
Agrigenetics Corporation
Advanced Research Division
5649 E. Buckeye Rd.
Madison, WI 53716 USA

Sabeeha Merchant
The Biological Laboratories
Harvard University
Cambridge, MA 02138, USA

Jan A. Miernyk
Biochemistry Department
322A Chemistry Bldg.
University of Missouri
Columbia, MO 65211, USA

Isaac H. Miller, Jr.
Bennett College
Greensboro, NC 27401-3239, USA

Leonard E. Mortenson
Exxon Research & Engineering Co.
U.S. 22 East
Clinton Township
Annandale, NJ 08801

Marcia A. Murry
Cabot Foundation
Harvard University
Petersham, MA 01366, USA

Ramaiah Nagaraja
CLSC 309
University of Chicago
920 E. 58th St.
Chicago, IL 60637, USA

Judy Nedoff
Agronomy Department
University of Wisconsin-Madison
Madison, WI 53706, USA

Daniel R. Nelson
Monsanto Company
700 Chesterfield Village
Chesterfield, MO 63017, USA

Kristin Nelson
100 E. Tyrone Road, #12
Oak Ridge, TN 37830, USA

Eldon H. Newcomb
Department of Botany
University of Wisconsin-Madison
Madison, WI 53706, USA

J. W. Newton
USDA
Northern Regional Research
Center
1815 N. University Street
Peoria, IL 61604

Carlos A. Neyra
Department of Biochemistry
and Microbiology
Rutgers University
Cook College
New Brunswick, NJ 08903, USA

Dale Noel
Department of Biology
Marquette University
Milwaukee, WI 53233, USA

Stefan Nordlund
Department of Biochemistry
Arrheniuslaboratory
University of Stockholm
S-106 91 Stockholm, SWEDEN

Patricia Novak
Johns Hopkins University
Department of Biology
3400 N. Charles St.
Baltimore, MD 21218, USA

Mark R. O'Brian
Department of Biology
The Johns Hopkins University
Baltimore, MD 21218, USA

W. L. Ogren
Department of Agronomy
USDA/SEA
1102 S. Goodwin Ave.
Urbana, IL 61801, USA

Yaacov Okon
Department of Plant Pathology
and Microbiology
Faculty of Agriculture
Rehovot, ISRAEL

Marion O'Leary
Departments of Chemistry and
Biochemistry
University of Wisconsin-Madison
Madison, WI 53706, USA

Lee C. Olson
Christopher Newport College
50 Shoe Lane
Newport News, VA 23606, USA

W. H. Orme-Johnson
Department of Chemistry
Massachusetts Institute of
Technology
77 Massachusetts Avenue
Cambridge, MA 02139, USA

C. B. Osmond
Biological Sciences Center
Desert Research Institute
Box 60220
Reno, NV 89506, USA

Christian Paech
Station Biochemistry
South Dakota State University
Box 2170
Brookings, SD 57007-1217, USA

Pedro Antonio Arraes Pereira
Department of Horticulture
University of Wisconsin-Madison
Madison, WI 53706, USA

Gerald A. Peters
C. F. Kettering Research Laboratory
150 E. South College Street
Yellow Springs, OH 45387, USA

Jay B. Peterson
Botany Department
353 Bessey Hall
Iowa State University
Ames, IA 50010

Richard B. Peterson
Department of Biochemistry and
Genetics
Connecticut Agricultural Experiment
Station
123 Huntington St., P.O. Box 1106
New Haven, CT 06504, USA

Donald A. Phillips
Department of Agronomy
University of California
Davis, CA 95616, USA

Jan Pitas
Agrigenetics Corporation
Advanced Research Division
5649 E. Buckeye Rd.
Madison, WI 53716, USA

Mark Pope
Department of Biochemistry
University of Wisconsin-Madison
Madison, WI 53706, USA

John Postgate
ARC/UNF
University of Sussex
Brighton, Sussex
UNITED KINGDOM BN1 9RQ

J. Prabakaran
CASAM-Microbiology
Tamil Nadu Agricultural
University
Coimbatore 641003
Tamil Nadu, INDIA

Glenn G. Preston
Department of Biochemistry
322A Chemistry Building
University of Missouri
Columbia, MO 65211, USA

Laura S. Privalle
Department of Biochemistry
Duke University Medical Center
235 Nanaline H. Duke Bldg.
Durham, NC 27710, USA

Velupillai Puvanesarajah
Departments of Microbiology
and Chemistry
The University of Tennessee
Knoxville, TN 37996, USA

Anton Quispel
Department of Plant Molecular
Biology
Botanical Laboratory
Nonnensteeg 3
2311 V. J. Leiden
THE NETHERLANDS

Douglas D. Randall
Department of Biochemistry
322 New Chemistry Building
University of Missouri
Columbia, MO 65211, USA

Linda Jo Rasmussen
Department of Biochemistry
University of Wisconsin-Madison
Madison, WI 53706, USA

Margaret Redinbaugh
Chemistry Department
SUNY-CESF
Syracuse, NY 13210, USA

Paul Reibach
Department of Agronomy
and Plant Genetics
University of Minnesota
St. Paul, MN 55108, USA

Minocher Reporter
C.F. Kettering Research Laboratory
150 E. South College St.
Yellow Springs, OH 45387, USA

Francoise M. Robert
Department of Soil Science
University of Minnesota
1529 Gortner Avenue
St. Paul, MN 55108, USA

Gary Paul Roberts
170 Belvidere Ave.
Fanwood, NJ 07023, USA

Paul Robin
Labo. Biochimie et
Physiologie Vegetale
Institut National de la
Recherche Agronomique
Place Viala
34060 Montpellier, FRANCE

Juan Carlos Rosas
Horticulture Department
University of Wisconsin-Madison
Madison, WI 53706, USA

Tomas Ruiz-Argueso
Department of Microbiology
E.T.S. de Ingenieros Agronomos
Madrid 3, SPAIN

Leonard L. Saari
Department of Biochemistry
University of Wisconsin-Madison
Madison, WI 53706, USA

Lakshmi Sadasivan
Department of Biochemistry
and Microbiology
Cook College
Rutgers University
New Brunswick, NJ 08903, USA

Samir H. Salem
Botany Department (Microbiology)
Faculty of Agriculture
Zagazig University
Zagazig, EGYPT

Seppo O. Salminen
Department of Agronomy
OARDC
Wooster, OH 44691, USA

G. Sarath
Department of Agricultural
Biochemistry
University of Nebraska-Lincoln
Lincoln, NE 68583-0718, USA

G. R. K. Sastry
Department of Genetics
University of Leeds
Leeds, ENGLAND LS2 9JT

S. C. Schank
Department of Agronomy
University of Florida
Gainesville, FL 32601, USA

Maria G. Schell
9916 Coluzi Drive
Knoxville, TN 37923, USA

H.-L. Schmidt
Lehrstuhl fur Allgemeine Chemie
und Biochemie der Technischen
Universitat Munchen
8050 Freising-Weihenstephan
FEDERAL REPUBLIC OF GERMANY

Lawrence Schrader
Department of Agronomy
University of Wisconsin-Madison
Madison, WI 53706, USA

J. P. Schumann
Cold Spring Harbor Laboratory
P.O. Box 100
Cold Spring Harbor, NY 11724, USA

Bruce R. Selman
Department of Biochemistry
University of Wisconsin-Madison
Madison, WI 53706, USA

Vinod K. Shah
Department of Bacteriology
University of Wisconsin-Madison
Madison, WI 53706, USA

Robert Shatters
Botany Department
Eastern Illinois University
Charleston, IL 61920, USA

John E. Sherwood
Department of Microbiology
and Public Health
Michigan State University
East Lansing, MI 48824, USA

Y. I. Shethna
Department of Biochemistry
University of Wisconsin-Madison
Madison, WI 53706, USA

Frank Simpson
Department of Biochemistry
University of Wisconsin-Madison
Madison, WI 53706, USA

Mahavir Singh
Department of Genetics
University of Bayreuth
Postfach 3008
8580 Bayreuth
FEDERAL REPUBLIC OF GERMANY

Rama Kant Singh
Microbiology Program
210 S. Frear Laboratory
The Pennsylvania State University
University Park, PA 16802, USA

John Smarrelli
Department of Biology
Loyola University of Chicago
6525 N. Sheridan
Chicago, IL 60076

M. T. Smith
Department of Biochemistry
322A New Chemistry Building
University of Missouri
Columbia, MO 65211, USA

R. Stewart Smith
The Nitragin Company, Inc.
3101 W. Custer Avenue
Milwaukee, WI 53209, USA

Russell L. Smith
Marine Science Institute
The University of Texas at
Austin
P.O. Box 1267
Port Aransas, TX 78373, USA

Doug Smyth
Department of Biochemistry
University of Georgia
Athens, GA 30602, USA

Seung-Dal Song
Department of Biochemistry
University of Wisconsin-Madison
Madison, WI 53706, USA

Thomas K. Soulen
Botany Department
University of Minnesota
St. Paul, MN 55108, USA

Martin Spalding
Plant Research Laboratory
MSU/DOE
Michigan State University
East Lansing, MI 48824, USA

Gary Stacey
Department of Microbiology
The University of Tennessee
Knoxville, TN 37996-0845, USA

S. Edward Stevens, Jr.
Department of Biochemistry,
Microbiology, and Molecular
and Cell Biology
The Pennsylvania State University
University Park, PA 16802, USA

W. D. P. Stewart
Department of Biological
Sciences
University of Dundee
Dundee DD1 4HN, Scotland
UNITED KINGDOM

John G. Streeter
Department of Agronomy
Ohio State University
OARDC
Wooster, OH 44691, USA

Larry W. Stults
Department of Biology
The Johns Hopkins University
Baltimore, MD 21218

Robert E. Stutz
25310 Elena Road
Los Altos Hills, CA 94022

Fumiko Suzuki
Department of Biochemistry
322-A Chemistry Building
University of Missouri
Columbia, MO 65211, USA

Shigeyuki Tajima
Laboratory of Biochemistry
Kagawa University
Miki-cho, Kiga-gun
Kagawa 761-07, JAPAN

Shiv R. Tandon
Department of Biology
University of Wisconsin
Platteville, WI 53818, USA

Elisha Tel-or
Department of Agricultural
Botany
Hebrew University of Jerusalem
P.O. Box 12
Rehovot 76100, ISRAEL

Irwin P. Ting
Department of Botany and
Plant Sciences
University of California
Riverside, CA 92521, USA

Louis S. Tisa
Department of Bacteriology
University of Wisconsin-Madison
Madison, WI 53706, USA

John D. Tjepkema
Department of Botany and
Plant Pathology
University of Maine at Orono
Orono, ME 04469, USA

N. E. Tolbert
Department of Biochemistry
Michigan State University
East Lansing, MI 48824, USA

Eric E. Triplett
Department of Plant Pathology
University of California
Riverside, CA 92521, USA

William T. Tucker
Research School of Biological
Sciences
Australian National University
P.O. Box 475
Canberra City, ACT 2601
AUSTRALIA

Raymond E. Tully
Department of Plant Pathology
and Crop Physiology
302 Life Sciences Building
Louisiana State University
Baton Rouge, LA 70803

Rodolfo A. Ugalde
Department of Bacteriology
University of Wisconsin-Madison
Madison, WI 53706, USA

Pat J. Unkefer
Agronomy Department
University of Wisconsin-Madison
Madison, WI 53706, USA

Robert G. Upchurch
Syracuse Research Laboratory
Allied Corporation
P.O. Box 6, Milton Ave.
Solvay, NY 13209, USA

Carroll P. Vance
Department of Agronomy and
Plant Genetics
The University of minnesota
1509 Gortner Ave.
St. Paul, MN 55108

Kathryn A. VandenBosch
Department of Botany
University of Wisconsin-Madison
Madison, WI 53706, USA

Larry N. Vanderhoef
Division of Agricultural
and Life Sciences
University of Maryland
1104 Symons Hall
College Park, MD 20742, USA

Donald VanderVelden
Department of Biochemistry
University of Wisconsin-Madison
Madison, WI 53706, USA

Silvio Viteri
Soil Science Department
University of Minnesota
1529 Gortner Ave.
St. Paul, MN 55108, USA

Thomas J. Wacek
Kalo Inoculant Laboratories
1145 Chesapeake Ave.
Columbus, OH 43212, USA

Daniel B. Wacks
Department of Bacteriology
University of Wisconsin-Madison
Madison, WI 53706, USA

Judy D. Wall
Department of Biochemistry
322 Chemistry Building
University of Missouri-Columbia
Columbia, MO 65211, USA

S. P. Wani
Department of Microbiology
ICRISAT Patancheru P.O.
Andhra Pradesh 502324
INDIA

James Waters
Department of Biochemistry
322A Chemistry Building
University of Missouri
Columbia, MO 65211, USA

Margaret Werner-Washburne
Department of Botany
University of Wisconsin-Madison
Madison, WI 53706, USA

Sarah E. Willis
Agrigenetics Corporation
Advanced Research Division
5649 E. Buckeye Rd.
Madison, WI 53703, USA

Lawrence J. Winship
School of Natural Science
Hampshire College
Amherst, MA 01002, USA

Harry C. Winter
Department of Biological Chemistry
The University of Michigan
Ann Arbor, MI 48109, USA

Jonathan Wittenberg
Department of Physiology
and Biophysics
Albert Einstein College of
Medicine
Bronx, NY 10461, USA

Carolyn A. Zeiher
Department of Biochemistry
322-A Chemistry Building
University of Missouri
Columbia, MO 65211, USA

Israel Zelitch
Department of Biochemistry
and Genetics
Connecticut Agricultural
Experiment Station
New Haven, CT 06504, USA

PROPERTIES OF A TREHALASE IN
CRUDE CELL EXTRACTS OF

OF GLUTAMATE AND OTHER
SPECIES

PLANT CARBON-NITROGEN
THE ROLE OF PYRUVATE

SITE SPECIFIC

E. B. Fred, P. W. Wilson, R. H. Burris, R. B. Peterson (photo taken in 1978 by Steve Dunbar)

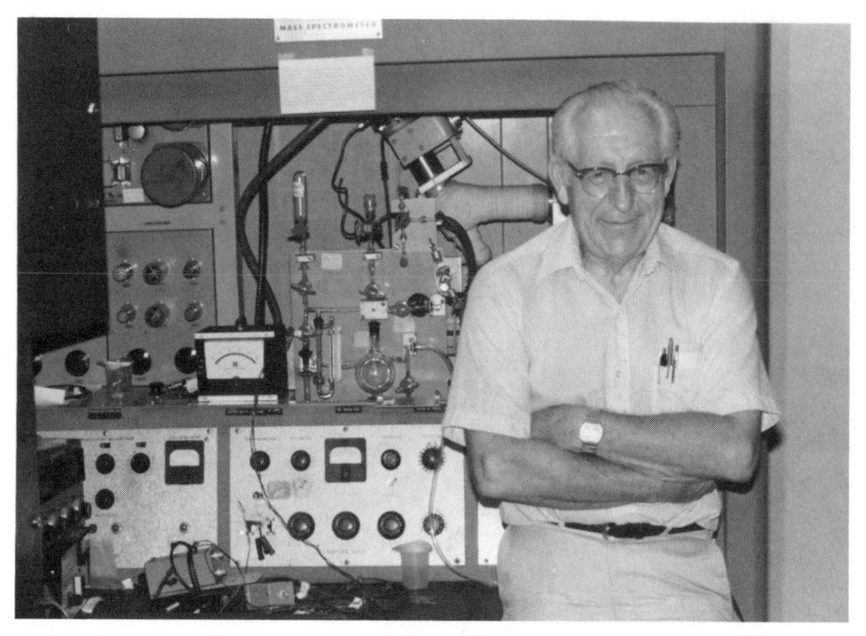

R. H. Burris with Consolidated Van Nier Isotope Ratio Mass Spectrometer

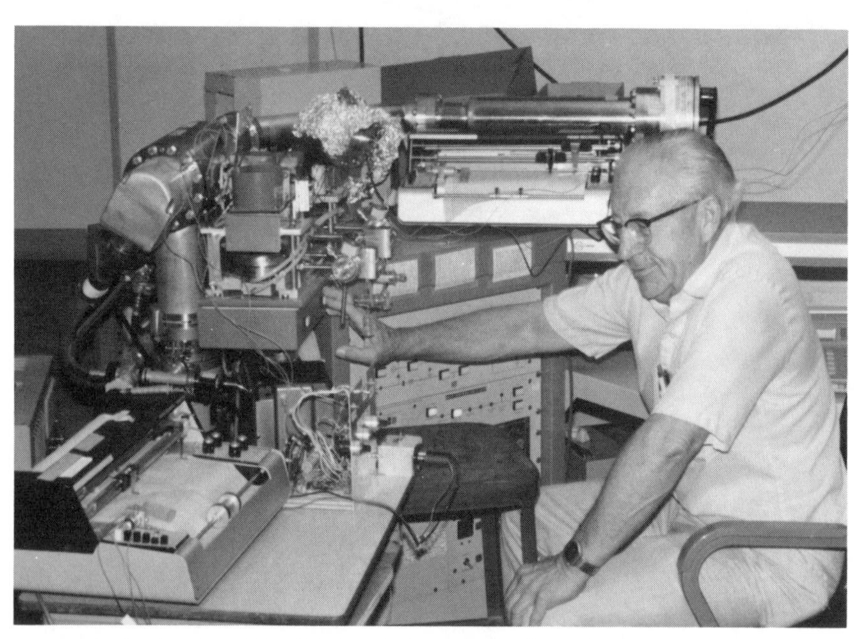

R. H. Burris with MAT-250 Isotope Ratio Mass Spectrometer

THE BIOCHEMISTRY OF THE LEGUME NODULE

The Whistling

REBECCA NETLEY

MICHAEL JOSEPH

UK | USA | Canada | Ireland | Australia
India | New Zealand | South Africa

Michael Joseph is part of the Penguin Random House group of companies
whose addresses can be found at global.penguinrandomhouse.com

First published 2021

002

Set in 13.75/16.25 pt Garamond MT Std
Typeset by Jouve (UK), Milton Keynes
Printed and bound in Great Britain by Clays Ltd, Elcograf S.p.A.

The authorized representative in the EEA is Penguin Random House Ireland,
Morrison Chambers, 32 Nassau Street, Dublin D02 YH68

A CIP catalogue record for this book is available from the British Library

HARDBACK ISBN: 978–0–241–53399–4
TRADE PAPERBACK ISBN: 978–0–241–53400–7

www.greenpenguin.co.uk

For my husband, Hugh
and beautiful sons,
Thomas and James.

Stealthy as the winter frost, it found
a rip upon the air.
And slipped from death to walk the night
But left no footstep there.

Anon

PART ONE

I

1860

I was one of only five passengers taking the boat to Skelthsea. There were no friends or relatives to wave me off and little comfort to be found inside my good wool coat. Wind blew in from a sea that was the grey of lead and the scream of gulls was harsh on the morning lull. As the boat rounded the rocks and approached the jetty, I picked up my cases and followed the others down the rutted pier.

I was helped on deck, where it began to drizzle; a thin, spitting squall, icier than I could have imagined, and I steadied myself with the rail as the vessel tipped this way and that. My fellow travellers showed scant interest in me other than a passing glance, and why should they? I have the sort of face and figure that is unremarkable – a complexion without bloom and eyes shadowed by sleepless nights.

The final passenger, a woman in her early middle age, smiled at me once before disappearing below.

Deckhands lugged boxes of supplies from the jetty

and it was another forty-five minutes before the steamboat left the harbour. As if on cue, the rain fell in earnest, pitting the sea.

'You'd do well to go down, out of the weather, missy,' the captain said.

But although gusts lashed at my cheeks, I wanted to hold that moment a fragment longer and did not seek the cabin until a sea mist fell and the land was finally lost from view. Until then, I believe I had not fully realized my decision. Even as I had packed and laid my clothes carefully in the cases, it had seemed that I was in some dream. My friend had watched me from the bedroom door, and shaken her head with something like pity.

'Are you sure it's wise, Elspeth?' she had pressed. 'To leave all your friends behind? The only life you know?'

But I had held tight to the letter – and to the way it lit a candle in a dark place. My friend had sighed. And meanwhile, I envisaged Skelthsea, its crags dotted with gulls, how waves might crash and break on the rocks, and a tide that washed the sands endlessly new.

My fingers went to the locket which held my sister's image. Of all my losses, it was hers that I thought of most, and as we drew further from the mainland the pain came again and stung as hard as it had in those early days.

I scarcely recall the journey, only that, as we neared the island, I climbed from the cabin to watch our approach.

I had been told that Skelthsea was beautiful but the reality stole my breath. Here was the hulk of cliff, the sweep of valley and, high up, a curving ridge that caught splinters of the dying sun. The beach was crowded with people waiting for the boat's arrival, chests and parcels ready for transport back to the mainland.

Sitting atop a hill lay a house larger than the others. Smoke rose and wound about the gabled roof. I surmised that this must be Iskar, the place that would now be my home. Leaded windows framed many rooms and with the setting sun behind it, the house cast a long shadow beneath which gorse and scrub shivered in the autumn chill. I tried for a smile but my face was frozen. The boat drew up to the pier and the engine fell to silence.

'Are you Miss Swansome, the new nanny?' A man came aboard and took my cases. He did not smile.

I nodded and, without a word of introduction, he carried my bags from the boat and waited as I was helped on to the barnacled pier. It was warmer on land but afternoon had begun to pool in the curve of the bay. The villagers' attention turned to me with nods and tentative looks whilst the children regarded me with open curiosity. In spite of the temperature, most were barefoot.

'This way.' The man walked ahead.

I tried to keep up, my feet slipping on shingle as we began the steep path to Iskar's grounds. Rose bushes blackened by frost grew wild and spilled over crumbling walls. Borders that had once, no doubt, been colourful and abundant had been reclaimed by sea grass and saplings and I became conscious of the smells – fish and peat smoke and a thread of sweetness: heather or gorse.

'Have you lived here long?' I asked.

'All my days.' And his eyes wandered to the beach, where fingers of orange were spreading across the horizon.

'And are you in service to Miss Gillies?'

'Aye, since I was thirteen.'

Birds circled the ridge and I looked up; a woman stood, curly hair and skirt flying. For a few moments, I felt her eyes upon me, and then my companion's feet began on the path once again.

Rounding a bend, we came to the house itself and I could see now the rot in the timbers and the crumbling of stone. The carvings on the gabled door were worn nearly flat like ink lines on an illustration.

Once inside, he placed my luggage in the hall, home

to a smouldering fire, and nodded to one of the two straight-backed chairs that framed the hearth. I duly sat and his boots rang out along the corridor until the closing of a door signalled that he had gone.

I took in a dull gleam on the furniture and the hovering of dust. Stags' heads with blind gazes hung above hunting prints and the flickering lamps dropped pools of yellow light to the floor. On one of the walls was a small portrait. I got up and saw that it was a painting of Mary, the child I had come to look after. Here she was, a little younger than in the photograph I had been sent.

Beside her stood her twin brother. A face I was destined never to meet.

As I studied her features, I recalled the contents of the letter; she, like me, had lost all those closest to her. I listened then to the immense quiet, imagined Mary in the web of rooms, left to the mercy of her aunt. Her aunt and now me. I felt then not just the strangeness of the unfamiliar house but something else, a quality to the quietness that seemed unnatural, and experienced the tiniest nibble of some doubt.

The silence was broken by footsteps.

'Miss Swansome?' The woman who came into view had a broad island accent and the first full smile I had received that day. 'I'm Mrs Lenister, I keep house here.' Her wiry hair was held tightly beneath a cap and she had bird-like eyes that were sunk in the lines of her face. But for all that, her eyes were kind and a pretty shade of blue.

'Trust Angus to leave you all alone here. I bet he didn't even introduce himself, am I right?' And I felt my edges begin to thaw as she beckoned me along one of the long halls, where our shoes echoed on the tiles. A smell of meat and baking drifted on the passageway and I realized that I had not eaten since a small breakfast.

As we walked, she kept up a constant stream of chatter.

'Miss Gillies is sorry not to be here to meet you, but she had urgent business with the tenants and Mary went with her. She's hoping to see you later, if you're not too tired. Meanwhile, she instructed me to give you tea and some food after your journey. It's warmer here if you're happy to take a seat with me in the kitchen?'

I felt a pinch of disappointment. I was anxious to see for myself what sort of girl I would have the responsibility of. But I said that I considered it no hardship; I knew from my own home that on cold days, it was most likely the warmest place in the house.

The kitchen was vast, with marbled sinks, a huge range, pots and pans hanging from racks and in the centre a table, scarred many times with the scores of careless knives. Salted fish and dried herbs hung from hooks on the ceiling.

The teapot gave off a faint wisp of steam as she poured and pushed a bowl of stew and a plate of cold fare in my direction. I ate gratefully, listening with only half an ear to her conversation.

'It's good to have you. Mary's in proper need of a nanny.'

'How long since the last one left?'

'Hettie?' Her lips tightened a fraction. 'A few months now, but it's not always easy to find a replacement. There's not many would want to leave the bustle of an Edinburgh life for the likes of here.' Her eyes raked me curiously.

'Hettie chose a bad time to leave,' I said. 'Mary must still have been grieving for her mother.'

'She was. They both were.'

'I was very sorry to hear that William had died. How long since they lost him?'

'Barely weeks after Hettie had gone.'

'Miss Gillies did not say in her correspondence how he died.'

Mrs Lenister pulled a chopping board towards her and began peeling apples. There was an awkwardness to her manner. 'Did she not? Well, it was an accident, but the mistress prefers we don't discuss William, I'm afraid.'

I wondered at that. After Papa had passed, Clara and I remembered him often in our conversations; it brought pain but there had always been comfort too. Perhaps here, the subject of William's death caused too much distress.

'You look concerned, Miss Swansome,' she said. 'Your post here should be an easy one. Mary is an obedient child.'

I decided I had questioned Mrs Lenister enough. My eyes took in some of the details of the kitchen – the windows and walls running with condensation and a dark patch on the ceiling which grew green spores at its edges.

While I drank my tea, we talked of my journey and a little of where I had lived in Edinburgh. When I had finished, she suggested she show me the way to my bedroom.

Walking a little in front, she guided me through the house and up a staircase to where a mullioned window captured prisms of the autumnal rays. As we went, Mrs Lenister enlightened me as to the geography of the house. The walls were panelled and the carpeting so thin as to show through in places. I had imagined more: a residence with plusher furnishings, brightly lit and welcoming.

Although oil lamps stood on tables, their hesitant light barely licked at the gloom. It was tomb silent, as if even the air did not dare make a sound, and with each step I imagined the twins' laughter flying down the corridors and ringing in the stairwells, their faces flushed, figures racing in a game of chase. And all the time I thought of the boy who was no longer alive, who lay in some fresh grave upon the island.

The hush was profound, as if weighted with shock.

Finally, we came to the end of the hall, where ahead, a narrow staircase led up to another floor.

'What's up there?' I asked.

'They were the old nurseries. It's where Hettie and the children slept.' She made a vague gesture with her hand and then paused for a fraction. 'We moved everything downstairs after she left.'

'And where is William's room now? Has it already been packed away?'

'William slept in that wing,' she said, indicating a door. 'It's kept locked now.'

'Locked?' I echoed.

'It's an old building, the floors are beginning to rot in places,' she said by way of explanation, 'but you are here and must be keen to rest.' She led me to the room opposite.

She must have caught my expression of shock and shook her head. 'That was not the cause of his accident, Miss Swansome.' But it was clear from her look that she was keen not to say more.

In my bedchamber, the surfaces shone and peat was stacked neatly for a fire, my cases already placed at the foot of the bed.

'Greer has done a nice job for you here,' she said, and I was overcome suddenly with exhaustion.

'Thank you,' I said, 'and thank you for the tea.'

She hesitated, and after a beat, she came in further, closing the door behind her.

'Miss Mary,' she picked her words carefully. 'She's been through an awful lot. Did Miss Gillies tell you? She's not spoken since her brother's death.'

'What do you mean?'

'Not a word. Not one word since he died.'

'She's mute?' I felt a stab of dismay. Miss Gillies had not mentioned this.

I could sense her awkwardness. 'But don't take her silence personally; it is how she is with everyone.'

'I imagine she's still grieving.'

'That's it exactly. Exactly,' Mrs Lenister said with relief. And, as if satisfied, she tucked her apron tighter and gave me a quick smile. 'I'm downstairs if you want me. We eat at seven.'

When she had gone, I sat on the bed, limbs weighty with fatigue, and tried to anchor myself to the moment; but my skin felt half-empty and I thought of Edinburgh. I did not know any more who or what I was, only that I had left behind, forever, the person I had once been.

I stayed where I was in the half-light that heralded the passing of the day and breathed in the strange new air of Iskar.

From my bag, I took out the photograph that Miss Gillies had sent – a little curled at the edges – and peered into its image. The picture was of the twins and Hettie; none smiled for the camera. I had spent so much time over the past weeks in examination of their faces that I knew them by heart, but it was to Mary's that I went again and tried to paint the information that I now held on to her features.

My gaze went to the mouth where her silence gave it new shape. Behind them stood Hettie, her eyes dark and intelligent, a twist of hair falling against a cheekbone. Peering closer, I saw that the surface of the picture was spoiled – that some flaw in the development had left William's likeness slightly darker and more blurred than his sister's.

I shivered – it was as if the photo itself foreshadowed the coming tragedy.

2

I was woken the next morning by the sound of waves rushing against shingle and the cry of gulls. Sun slipped in through a gap in the curtains. It took a few moments to orientate myself but checking the clock, I leapt from the bed, anxious that I had slept through and failed to emerge for supper. I washed and dressed quickly and made my way to the hall, where a maid I hadn't seen before hummed quietly as she worked.

'Good morning. Which is the dining room, please?' I asked although the aroma of coffee already signalled the direction.

She ceased her tune and looked me up and down without a smile. 'This way.'

Downstairs, Mrs Lenister was already dusting crumbs from the table. 'I hope you slept well, Miss Swansome. Miss Gillies said that she will see you in the drawing room after breakfast.'

'I didn't mean to sleep through –' I began, but she waved a dismissive hand.

'Miss Gillies is always tired after visiting day; she

said it was a blessing that you could both meet fresh this morning.'

I helped myself to bread and fish. The coffee was hot and welcome and I ate and drank quickly, keen to meet my new employer.

When I had finished, I brushed nervously at my skirts and knocked at the drawing room door.

She called for me to enter and I stepped inside. The corridor had been chilly, but here the fire warmed the air and caught in its glow the polished edges of occasional tables and elegant desks. Two large windows let in the autumn light, and my eye was drawn to a glass case of brightly coloured birds standing on a lacquered table.

Violet Gillies sat in a chair with her right cheek to me. She wore a black dress with silver buttons. Pearl-headed pins glinted in her hair which was teased into curls at the nape.

Although she knew that I stood there, she took a while to turn and, when she did, I had to hide my shock. In profile, there was nothing to hint at the scars that she wore on her other cheek.

I was not quick enough to conceal my reaction and for the briefest of moments, something in her irises flickered – displeasure or pity, I could not tell which. She rose and motioned for me to sit.

'It's lovely to meet you at last, Elspeth.'

She must have been in her early to mid-thirties with clear eyes and a high, smooth forehead.

With a finger, she touched the burned skin. 'This is something I have lived with since my sixteenth birthday. I'm accustomed to the response it brings.'

'I'm sorry,' I said.

'It was an accident.' There was compassion in her tone. She knew the circumstances that had led to my taking the position – the tragedies that had chased me into my own particular corner.

'I'm sorry I was not able to meet you yesterday. I hope you were comfortable last night.'

'Thank you, yes.'

She laid her ringed hands upon her lap as if punctuating the end of her enquiries. 'I can't tell you how welcome you are. It's been a demanding time. Mary's mother, Evangeline, passing early this year, and then William. Their father died a long while ago – there are so few of us left now.'

'That is very sad.'

'Sad and difficult. I'm not used to children and their ways. We've managed somewhat between us since Hettie left, but I have been keen to fill her position. As I explained in my correspondence, you will find us more informal than on the mainland and we have reduced ourselves to using only as much of the house as we need. All the same, we manage nicely.' Although her tone was assured, there was a look of entreaty in the tight anticipation of her features.

'I'm happy to be here,' I said.

'At nine years, I know Mary is a little old for a nanny, but her specific needs demand it.' She picked at the edges of the chair. 'I'm sorry to have to tell you now – perhaps I should have let you know in my letters – but she has not spoken since the day she lost her brother. I had hoped by the time you arrived that might have changed, but regrettably no.'

'Not one word?' I did not admit that I was already in possession of this information.

'Not a syllable. It has been a heavy trial for all of us.' I could feel the concentration of her focus and I slightly adjusted my impression. Violet Gillies had steel in her marrow.

'But otherwise, how is she faring?' I asked.

'Mary's taken her brother's death badly but it's temporary, I'm sure. There are occasional nightmares and bouts of sleepwalking, but that is in keeping with her loss, so I'm informed, and will pass in time with your kind attention.

'I still find it hard to take in myself – that such things can happen so quickly – but you are particularly well placed to offer comfort on that score as you have suffered similarly, have you not?' She eyed me curiously.

'I lost my mother and brother some years ago, my father in the last two and,' for a moment I stumbled on my words, 'my sister only recently.'

'You took care of your sister after your father passed?'

'I always had the care of her.'

'And there was a fire?'

'Unfortunately.' I looked away to hide the bolt of pain.

She nodded sadly but there was a flash of relief. 'We are the same, then. Life can be so cruel, but you must consider Iskar your new home and myself and Mary your new family. I understood from your correspondence that, before she died, your sister had difficulties of her own?'

I did not want to revisit Clara's face, so pale and round, or the lively gaze that was sometimes so piercing. 'Clara was never very well. She was called slow by some because she was clumsy with her movements, but her mind was quick.' With a pang, I recalled the witty retorts and the way her thin fingers struggled stubbornly over a button. 'Her birth had been difficult,' I said by way of explanation.

Miss Gillies' shoulders relaxed and a shaft of sun caught the feathers of the birds in the case, giving them life. For a moment, I could imagine them lifting their wings in flight. 'I think you are well suited to the position, Elspeth. I believe that you'll fit us very well.' She smiled, showing a row of even teeth.

I could feel the weight of expectation; the pressure of the house around me, the air trapped in corridors above. And beyond that, somewhere in this maze of rooms, a child whose mouth captured only silences.

'And to Mary,' she continued, 'she will give you no hardship, I promise. From one to four I will give Mary lessons and you are free to do as you will.'

With such an easy workload, I wondered at Hettie's leaving at all and felt a sudden dislike for the girl with beautiful eyes.

'Why did Hettie leave?' I asked.

'My sister's death took her very hard.' She leaned over and adjusted the fire then changed the subject before I could press further. 'As to Mary. She is a girl who wants to laugh again, a girl who loves to walk on the beaches and see the seals and the gulls.' Her eyes wandered as if recalling her at an earlier time. 'Don't be fooled, however.' She leaned forward. 'She does not talk but she is very clever. Very clever indeed.' And although she spoke with pride, her face shrank back on it.

There was a pause in the air as if the atmosphere had stretched a little around me.

Miss Gillies' eyes fixed on the window. I followed her gaze and there, through the panes, stood Mary herself, hair bleached in the watery light, gazing back in at me with a strange and shuttered expression.

'And there she is,' Miss Gillies said and summoned her with a gesture that sent the bracelets at her wrists jangling. Now the moment had actually arrived, a film of sweat lay in my palm.

3

There was a polite knock and finally, Mary was before me – as like her image as a person could be and yet with all the qualities and deficits that real life endows. She was of slight build, with an oval face, fair colouring and regular features. If it were not for the wan, closed expression she wore, I would have called her pretty. At first sight, her dress seemed pristine, but then I noticed a mark on the lap as if she had recently spilt something and her braids were already fighting to be loosed from their bows.

When she stepped up and put out her hand in a polite how-do-you-do, I was struck with the emptiness of her regard. In front of her, she clutched a dirty doll.

'And that is Bobbity,' said Miss Gillies with a note of distaste and Mary's fingers gripped the toy harder.

Kneeling, I said hello to Bobbity and then addressed Mary herself. 'I hope you and I will have lots of fun. I am relying on you to show me the island and the house.'

Mary said nothing. There was no change to her stiff features. If anything, I sensed a further withdrawal.

'There, see,' Miss Gillies said. 'I feel already you shall be the very best of friends.' But a quick pull of her lips made me suspect that she did not believe it. 'Take a walk on the sands today and get to know each other. Perhaps, just this once, I will allow Mary a few sweets.'

I looked out of the window to where the veiled sun now sat behind a swathe of copper-edged cloud. I was keen to see Skelthsea and its celebrated beauty in full daylight.

'Shall we walk to the shop in the village, Mary? Your aunt says you may have some sweets. Do you like sweets?'

'Mary, like all children, likes sweets and they are a rare treat. I don't generally hold with giving young ones such things.' Mary did not smile, but there was a flicker of something I moulded into pleasure.

Outside, wrapped in our coats, we walked into a gusting wind that blew in the shriek of gulls and the shushing of waves as the swell rose and receded on the beach. The light was brittle and uncertain, as though captured through glass, and I filled my lungs with the sharp air.

'Well, Mary,' I said, 'you must show me all the best

places for walking. When I was younger I went fossil hunting with my father. Have you ever done that?' Mary did not answer and after a time, I too fell quiet, casting her the occasional glance and digesting what I had so lately learned. Our feet crunched on the path.

On the beach, I paused to pick up a shell, pink and coiled like the inside of an ear. Mary stooped too, catching my eye, and together we foraged along the sand. As she crouched, I took the opportunity to observe her. The frock coat she wore came to her knees and below that her legs were long and thin.

I realized how inadequate my expectation had been. It was as if I had believed that I had never been destined to take charge of a real person, but only a doll to be formed to my own needs. In my imagination, I had assumed the task would be within easy reach of my skills, but faced with the real object of my purpose, my confidence was pricked. There hung about her an aura of solitude that touched me. I remembered how I too had retreated after Clara's death, to a place beyond comforting.

From time to time she glanced at me, a dash of interest in her look, and I studied her eyes, the almond shape that held the brown of her iris. I wondered what she saw when she regarded me and what she felt about my coming. If Hettie had left so quickly, she might not want to trust me for fear that I too might take flight.

After a while, we turned back and took the path that led to the crofts. The valley was hewn into patches of crop hemmed by walls. Highland cattle grazed the higher ground and sheep roamed freely. The village consisted of a row of about fifty cottages strung in an uneven line. The smell of smoke and fish and the faint noise of chatter grew stronger, and two girls came down carrying baskets; somewhere someone sung in Gaelic. Women sat in their doorways as they spun wool and chattered; I walked past them slowly with a good morning. They all wore headscarves of the same woven tweed and regarded me with a mixture of shyness and curiosity. Few answered my greeting.

The shop was larger than the other dwellings. A sign, rusted at the edges, announced it belonged to Reid Paterson. The interior was crammed with all manner of goods – from fishing, farm and household equipment to every variety of domestic fare. Nets and rods hung from the ceiling and I could smell the thickness of oil. Mary made her way straight to a counter, behind which stood a row of jars.

I picked up the bell and rang. A banging came from somewhere beyond and a man entered, stopping in his stride at the door to regard us. His skin was weathered but he was handsome in a rough kind of way, with dark hair and eyes. His coat was smart and his collar starched white. His eyes flicked over Mary with dislike or animosity.

'So, you're the new nanny at the big house.' He swept me from head to toe in a gaze that set me flushing and then he smiled with rude over-familiarity. I stood taller, pulling my coat tighter, and only nodded in reply.

Mary pointed to one of the jars and I bought the sweets, which he first weighed in a pair of filthy scales before pouring them into a cone of paper. As he worked, he tried to engage me in conversation, his eyes too intimate on my face and body. His manner put me ill at ease. I said little in return and soon he stopped trying.

I was relieved to be outside once again. We climbed the hill, my hair in its pins being tugged by the wind, and I had a desire to loosen my hat, release the clasps and feel it blowing around my face. The further we grew from Iskar, the lighter Mary's steps became, and she ran ahead to study a patch of gorse or to rest on a drystone wall and observe the cattle. And then she fell in beside me, and a tightness in my chest I had not been aware of loosened a little.

We reached the ridge and I had to lean forwards, hands on knees, to give my straining lungs time to ease. Turning, the whole valley lay before me, captured in the sun, a lucid green with cloud shadows chasing across the fields and racing up to the cliffs. It was beautiful. Far away, beyond the eye, lay the mainland.

Mary stood beside me, her dress blowing and the wind snapping at her boots, Bobbity in one fist. Tentatively, I reached down and took her other hand in mine. The warmth of her skin bled through my glove and, for a moment, I was back in Edinburgh, pushing through the crowds with the smell of burnt sugar and tobacco smoke, the sounds of the circus already clamouring in the air and Clara's palm clasped tightly in mine lest we become separated. A boy had passed by with a monkey on each arm and Clara had turned and laughed with delight.

When the recollection left me, standing cold and bereft on the ridge, I looked down at my charge.

Mary glanced back up and seemed to assess me and then, for the first time, the smallest curve of pleasure crept on to her lips.

4

Later, a lunch of soup followed by cheese and some sort of flat cake was served in the formal dining room with Miss Gillies herself. I became aware that she rarely addressed Mary, perhaps tired of receiving no answer or keen to develop our own acquaintance, but as the meal wore on her eyes seldom strayed Mary's way and I was bitten by faint perplexity.

'What do you think of our island?' Miss Gillies asked.

'I had not realized it would be so beautiful.' I could see she was gratified. 'Mary and I saw a sea eagle.'

'You know a little of these things?'

'My father was a keen naturalist. He taught us from a young age.'

'What was his profession?'

'He was a lawyer.' Too soon the smell of his pipe came to fill me with loss. I swallowed and put down my fork until the wave passed. I recalled Edinburgh and the doctor's assessing gaze, the quick glance he paid to my friend, how Mrs MacAllister had then clasped her hands in agitation. *'She will not eat. You must make her eat,'* she pleaded.

I shook the memory away. Although my throat was tight, I lifted the fork once more and put the food to my lips.

The minutes passed and at one o'clock, Miss Gillies took Mary for the promised lessons, leaving me free to wander and explore. I sat for a little longer by the fire as their steps receded and let the first morning come to rest. Although silent, I had not found Mary's company uncomfortable. It had not, I realized, been necessary to hold long, one-sided conversations. Companionship found other avenues: there had been some shared wonder in the foraging on the beach and in the sweet she offered me later. The morning had exceeded my expectation. Perhaps, after all, a child with problems would present not only greater challenges but greater rewards, and I felt a determination to change Mary's world for her if I could. I remembered my father's idiom: 'If you fail to reach the summit, climb another mountain.' I thought then of how, after Clara's death, I had so nearly given up, and I was grateful for the opportunity Iskar offered in giving me a path on a new mountain.

Eventually, I left the fire and began to explore, and as I trailed along corridors and peeped into rooms, I became more aware of the burgeoning decay – velvet curtains with patches worn thin, scuffed furniture and a smell of damp. I wondered how long those spaces had lain neglected. Below the panelling, there was dust where the oak had been bored by worm. On the upper floor, I came across another maid clearing out my grate. She blushed when she saw me, rising from her knees in an awkward curtsey. She had the darker, weathered skin and hair of the islanders.

'Good afternoon,' I said. 'I'm Miss Swansome, Mary's new nanny.'

She gave me a smile and bobbed. 'Magda,' she said in a pleasing, sing-song voice.

Ahead, darkness swelled in the garret stairwell and curiosity about Hettie made me pause at the bottom and then climb. The banister was worn smooth by the pressure of so many palms. Beyond the sounds of the island, another made itself heard: the faintest of whistles, like wind trapped in tiles or chimneys.

I stopped. Darkness bled into the corners and edged across the walls. As I listened, the whistling rose and fell, and I became aware of the way my hand tightened on the rail.

For no reason that I could fathom, there was a quality to what I heard that turned uneasily inside me.

At the top of the staircase, I opened a door into a largish bedroom, weighted with the sort of heavy furniture found elsewhere in the house. It was shrouded in darkness. I moved across to the window and drew the curtains, spilling light on to surfaces – a bed, a table, two armchairs and an imposing oak press. This must have been where Hettie slept, and I thought I detected the slightest scent of powder or perfume. On the dressing table a yellowed comb with missing teeth and a folded handkerchief sat in a film of dust. I imagined her at the mirror, examining her pretty face with satisfaction. It struck me that, wherever she was now, she did not know that William was dead.

I lifted the comb to the light where strands of red hair were caught in the bone. I reimagined her face and instead of the brown locks, saw auburn, catching the sun like embers. My features met me in the foxed glass and I put a finger to the disappointing shade of my own. There was no sparkle in the grey of my eyes, no blush on my thin cheek.

The drawers were empty but for a ripped stocking that had fallen to the back and caught on a knot of wood. I went to stand at the window where a layer of cold seeped in from the frame. The eaves above me creaked and I realized then how silent it was at the top of the house, sound muffled and absorbed by thick floors. Something in the air made me pause and I experienced a strange awareness of Hettie as if she stood just behind me, watching and judging. I envisaged her easing a shoe from her foot at the end of a long day or raising her arms to pull pins from her hair. Turning, the quietness seemed to shiver and I moved quickly back to the door.

The other rooms were neglected – a schoolroom where I tried to visualize William and Mary bent over desks with inky fingers, the lure of a blue sky in the panes beyond. Had they giggled behind their palms or passed secret notes between the pages of a book? The atmosphere still clung to some of the schoolroom smells of paint and industry.

As I returned along the narrow corridor, I tried not to compare the stillness and decay with the sunny rooms at home, but they lingered in my thoughts, pinching hard at my emotions. Clara had not taken lessons as her hands were too clumsy to write, but she had sat beside me when Miss Dodds came with her dusty maps and tomes on the kings and queens of England.

When I had wandered to my satisfaction, and with Mary still at lessons, I returned to the drawing room fire. At half past three, the maid I had heard humming that morning brought tea and bread and I studied her more closely. She was young, no older than myself, with thin, dark hair and a long face.

'Good afternoon,' I said.

Her eyes were dull with high, arched brows that framed a less than friendly expression. She bent her thick waist and placed the tea things on a table before me.

'Are you Greer?' I said. 'Was it you that made the preparations to my room?'

'That was me.'

'I thank you. It is much appreciated.'

Her gaze constantly shifted and there was an air of animosity about her.

I took up a piece of bread and put it on a plate. 'How long have you worked here?' I asked.

She paused, hands clasped behind her back, and fidgeted to be gone. 'I've been doing jobs here since I was nine or so.'

'You must know the house very well by now,' I said by way of conversation.

She gave the smallest of impatient sighs and her eyes slid to the door.

My curiosity about William would not be quelled, and in spite of her manner, I tried my chances. 'I was sad to learn of what happened to William. I have wondered what the accident was by which he died.'

Greer scowled and regarded me sullenly.

'It's getting colder,' I said, determined to draw a response.

She leaned over, placing the empty crockery on to the tray, and I got a waft of cleaning soda and beeswax. She had short fingers with square, roughly cut nails.

'They're tying up the boats so they're expecting a storm.'

Out of the window, fast-moving cloud skimmed the ridge. 'I thought I saw a pretty glass pane at the back of the house as I was coming up the path with Mary earlier.'

'That'll be the little chapel. The minister used to come and give services here, but the staff is too small now. It's past the library.'

After Greer had gone, I took to the corridor I believed led to the chapel. As I followed it, a low murmur of conversation seeped from one of the rooms and I stepped lightly, pausing outside.

'It's not for us to comment.' I heard Mrs Lenister's low, quick tone. 'You should do well to remember that.'

I pressed my ear closer, intrigued by what I heard.

'Miss Gillies has made it quite clear William is never to be talked of.'

'She's the sort of person who won't let things be. Questions, questions.' It was Greer's voice.

Whatever was said next was drowned out by the clattering of a peat bucket, but my instinct was right and Greer clearly found something about me to dislike. I wondered what had so early drawn her displeasure, but perhaps it was just jealousy of my elevated status or the fact that my occupancy would cause extra work. How poor she was at hiding her feelings, I thought, but I had dealt with surliness before and was determined to turn her with kind words and grace. As I made my way to the chapel I was struck with the content of what I had heard and I paused, wondering why Miss Gillies might be so reluctant to talk of her nephew's demise. But he was not long dead and grief must be the reason, I decided.

5

Later the household gathered in the hall for the journey to church. Greer had my coat ready and I took my place next to Mary who was wrapped in a muffler and hat. Greer, Mrs Lenister and Angus held lanterns by which we would make our way home. The servants left through the back entrance and met us outside.

The breeze had picked up, dislodging orchard leaves. Peat smoke sullied the air and the protesting sound of gulls cut through the rising wind, but I was glad of the wildness, loved how the island complained so loudly.

Villagers were leaving their cottages and proceeding in a wavy line of dark coats, headscarves and lamps. They waited politely for Miss Gillies to reach the street and then nodded and curtsied and followed behind. There was friendly chattering, along with raised caps and smiles. It was clear that Miss Gillies was held apart from the rest of the community. As we neared our destination, darkness had fallen and the circle of the moon showed itself in the night sky.

The Gillies pews were at the front of the chapel

and I followed the others, taking my place next to Mary. The murmuring gradually ceased and the silence was broken only with the occasional cough and the sound of benches creaking under adjustment. Somewhere behind me a baby mewled. Candles threw flickering shapes on to the whitewashed walls and a bat, caught in a high corner, lifted a wing and huddled down again like a chrysalis.

We knelt. Miss Gillies leaned forward for a long time, her head bent over her clasped hands, the nape of her neck peeking from the collar of her coat. Mary sat and stared ahead until Miss Gillies, with a tug on her sleeve, bade her to follow her example. I had not prayed since the day I returned home from visiting the Edinburgh family who wished me to tutor their daughters. I could still smell the smoke and remembered how, by the clenching of my heart, I had known that Clara had been caught in the flames.

When I rose from the floor, cold had already pierced my skin and my breath was white on the air. Candles shivered in the draught that passed down the aisle. The minister went quickly through a reading and then there was one hymn, complemented by the ringing of a tinny piano.

Time and temperature deterred a longer sitting and the service was over before it struck the half-hour. As we left, voices rose, relieved to have finished their evening worship and to return to fires, hot drinks and supper. The body of people moved slowly towards the door, pausing for conversation.

As I stood in line, I felt a gaze upon me. A tallish woman was watching me closely. We proceeded up the aisle until we came to her pew and were standing but inches apart. I gave her a smile which she did not return, but she continued to study me and there was something purposeful in her scrutiny, something beyond mere curiosity. It was only as I reached the exit that it came to me – there had been an expression of pity in her gaze.

Our party stopped to be greeted by the minister and Miss Gillies introduced him as Robert Argylle; next to him stood a woman I recognized from the boat.

'So this is Mary's new nanny.' The minister looked at me with interest. 'I hope you won't find our lives here too dull.' He had an angular face with sharp eyes that were shadowed as if by sleepless nights. 'And this is Mrs Argylle, my wife,' he said.

She leaned towards me, taking my hands in both of hers in warm welcome. 'I'm a very poor traveller,' she said, 'or I would certainly have made your acquaintance on the journey over. You must forgive me and let me give you tea very soon. We don't often have a new face on Skelthsea.'

Miss Gillies stood a little way off, holding Mary

tightly to her side, but nobody lingered long. The night had begun to settle and the wind had grown, sweeping down from the ridge and catching our hats and outerwear. We made our way quickly to the bay, where the sea's edge glimmered in the moonlight.

As we approached Iskar's path, my eyes drifted upwards to the nursery rooms and I stopped dead. Behind the glass there was a blur of movement. Miss Gillies turned, hearing the cessation of my feet on the shingle.

'Are you all right, Elspeth?' she asked.

I stared hard at the window, trying to make sense of what I had seen.

But there was nothing now to observe – and then a movement of cloud was reflected on the panes and I chided myself for my imagination, yet a feeling of unease remained.

Mary paused beside me and I turned to give her a smile but she, like myself, had her attention fixed on the garret rooms. I followed her gaze but was met with only a black veneer against the darkening sky.

In the hall, Mrs Lenister told me that dinner would be served in the dining room that evening with Miss Gillies and I wondered if usually I would be expected to take it alone with Mary. Coats and mufflers discarded, we ate roasted grouse while the storm lashed rain against the glass and howled in chimneys. The fires guttered and spat. Dinner was well cooked. Greer, who served us, seemed unconcerned by the weather, as did both Miss Gillies and Mary, and I gave an inward shrug.

Mrs Lenister came in with extra lamps, placing them on the sideboard, and we moved to a simple pudding of stewed fruit. All the while, Miss Gillies talked of the day she had spent, the lessons that had been completed with Mary, and questioned me about how I was finding my duties so far.

Occasionally, I turned to Mary with a comment but she ate stiffly, paying no attention to either myself or her aunt.

We took coffee in the drawing room where we played cards and I did not notice how late it was until the clock struck eight-thirty. We finished the round and Mary rose.

As we made our way along the upstairs corridor we met Greer, humming a tune as she checked the lamps.

'Good evening, Greer,' I said with my kindest smile but she only tightened her lips in answer before moving on.

In Mary's bedroom I got her ready for bed.

It was strange to be alone with her again and I was self-conscious. But her face was already becoming familiar and I realized that I liked it in spite of its quiet, inward gaze. There was something in the line of her cheek that recalled my sister – Mary's mouth, although sullen, lifted a little at the corners suggesting that once she had smiled and laughed much. Her hair was soft and carried a breath of the sea in its strands, and as I brushed through its length, I was soothed, remembering the nights I had done the same with Clara's – how I would glance up to meet her eyes in the mirror. Sometimes, I had wound her locks in rags and tied them and the next day we had combed them into curls.

'I can ringlet your hair, if you like,' I said but she did not react and so I plaited it into a loose braid.

When I had finished, she knelt at the bedside and prayed although she made no sound. Afterwards, I removed the copper warmer and pulled the bedding about her neck.

I hovered, wondering whether or not to kiss her, and all the while she looked up at me, Bobbity on the pillow beside her. I only had experience of being a sister, not a nanny, and I was unsure. Then I recalled the smile she had given me as we stood together on the ridge and I leaned over and placed my lips on her cool skin. Her lids were heavy as if she sought sleep and so, taking my candle, I left the bedroom, closing the door quietly behind me.

Later, I made my way to my room, which even a fire could not warm. When I snuffed the candle it was pitch-dark but for the flames that played on the walls.

As I thought about my day my mind drifted continually to Hettie and to the room above. I saw her catching red strands of hair in the comb and pulling through its thickness.

Every now and then the wind found some space or gutter to howl through and the shutters rattled.

I was starting to drift when I was half-woken by a hummed lullaby from the corridor outside, and I recalled Greer earlier, a tune on her lips. Her step was slow now as she made her way towards the room. When she reached my door, the volume fell nearly to silence, as if she did not want to wake me. The hush swelled uneasily. I imagined her turning down the lamps before making her way to bed. I waited for her to move on but there was a pause that lasted just a little too long and I felt the strange bristle of her dislike.

I crept deeper beneath the covers and listened until finally, her voice rose and then died as she drew further away, leaving me with only the snap of the fire to sing me to sleep.

6

The island grew colder, seeming to shrink against the coming of winter. As the days passed I became accustomed to the rhythms of the house and the people within. When I entered the drawing room or dining room now, I always went to the same seat, and when it was time for church, I was already anticipating the bite of the cold. Where we ate, be it with Miss Gillies or from a table in Mary's playroom, seemed ruled by whim. But the evenings were spent together, either in pursuit of a game or each engaged in our own occupations.

Every morning, Mary and I would venture into the wind and gradually, through the numbness of my emotions, sentiment began to find a path, awakening something in me that I thought had died with Clara. When I observed Mary, I believed that she too had begun to thaw to my presence: it was hard to quantify for she was neither demonstrative nor affectionate, but she was interested. Often, I would find her studying my face and once or twice, something I did or said brought an unwitting smile. But more than that,

I began to understand her better in relation to her life at Iskar and those within it – Miss Gillies' apathy, the lack of friends or playmates. Mary was not my sister but she was sorely in need of a loving one. And I tried hard to emulate that role, recalling all the tender things I did for Clara and passing them to Mary.

She still would not speak, but soon I began to notice the anticipation in her eyes at the start of the day and the way her gaze followed me if I got up from a chair or moved across the room to get something from Miss Gillies' workbox. These were things that gave me more pleasure than I had foreseen and I began to relax into my position and trust my footing. Yet despite the many things that brought me satisfaction, something in the background, just out of sight, picked at some obscure disquiet.

I avoided Greer, whose breath of dislike continued to follow me in spite of my attempts to garner her favour. One morning I came across her in the hall. She did not acknowledge me, but as she continued on her way she moved deliberately too close and her heel caught my foot, causing a flash of pain. I could not hide the flare of irritation.

'Be careful of your step,' I said, trying to keep the impatience from my tone.

She stopped and turned fully to face me, something

hardening in her gaze. After she had gone, I realized that she left me unaccountably a little afraid.

That day had been too wet to venture outside and Mary and I went to the playroom where I had suggested we draw each other's likeness.

Mary pinned her paper to the easel and I seated myself on a stool beside her.

'I shall draw you at your work,' I said. 'Let me look at your ears. Yes, they are very big. Like a cat's. I shall certainly make something of those.'

As I watched, her lips quivered, and in spite of her effort to remain quiet, she let out a laugh.

I regarded her with amazement and an answering joy. It was the first time that I had heard any sound come from her other than a breath.

I laughed too but she reddened and pressed her mouth closed in regret for her outburst. So I picked up my pencil and pretended not to have noticed and did not intrude on her discomfort.

As I drew, I looked up occasionally to study her. Now and then her hand went to touch some part of Bobbity, as if afraid she might have been snatched away. After a while, I concentrated on my own composition and let my pencil follow the lines of her face.

When Mary had finished her drawing she passed it to me – it was a portrait of myself with my short fringe and brown coat with the fox-fur collar. I was flanked by a boy and a girl. The girl was clearly Mary and I had to assume that the boy was William, but it was impossible to say because his face, unlike mine and Mary's, bore no features. I did not like the way the paper held on to that emptiness. Was this a response to Miss Gillies' curbing of all discussion about William? Or something else? I tried to read in Mary's expression some explanation but could find none.

Later, in my own room, I reflected on the absence of William's likeness. It might be the way she depicted death, and I could not argue – hadn't I too tried to deal with Clara's by rubbing the memory of her from me as much as I was able? Even so, it disturbed me and I laid it face-down beneath a hairbrush so that I did not have to witness it again.

The weather cleared and Miss Gillies told us that she would not be taking lessons that day as she had island business to conduct and so, after lunch, I suggested a walk in the garden. We explored the orchard and collected windfalls in a basket to take back to Mrs Lenister.

Although peat smoke from the island chimneys was comforting on the air, my consciousness was continually drawn back to Iskar where I searched its blank-faced windows and had some odd sense that I was watched. Turning to Mary, I saw that she too had fallen still and was looking up at the house. Although her expression was masked, something moved in her eye and I looked again, drawn by Hettie's room and the flat reflection of its panes. As we stood, a whistling sounded on the air and Mary clutched the doll tighter in her fingers, her lips paling. I opened my mouth to ask her if she had seen something or if she had heard it too and remembered that she would not speak, would not be able to explain even if she could. And so we continued, but only half absorbed in our pursuit as the wind rushed and pulled at the sea out on the bay.

When it was three by the clock we returned and found Mrs Argylle in conversation with Miss Gillies. They looked up as we entered and Mrs Argylle rose, giving me a warm smile.

'How are you, Miss Swansome, and you, Mary?'

Mary curtsied and took a seat a little way away and fiddled with the ribbons in Bobbity's tangle of hair.

'Mrs Argylle used to come once a week to give Mary and William lessons in French,' Miss Gillies said. She glanced at Mary sharply. 'Perhaps in time it can be resumed.' Her tone was severe and a spasm of anguish passed over Mary's features.

'I like to hope so,' Mrs Argylle said and gave Mary a look of compassion. 'Miss Swansome, I haven't forgotten my invitation. Perhaps if it's convenient you would like to walk back with me in a while? I'm sure Miss Gillies can spare you from Mary for an hour?'

Miss Gillies agreed and said that Mary could help with sorting out some threads and beads in her workbox, and soon I was collecting my coat and following Mrs Argylle back out into the cold. We spoke little on the way to the church as the words were blown from our mouths almost as soon as they were uttered, but it did not take long to reach the chapel, where their cottage sat at the rear.

The house was small but homely, smelling of something newly baked. A small dog that sat by the fire lifted its tail, gave a deep sigh and went promptly back to sleep.

A maid brought us tea and cake. 'So tell me, Miss Swansome, how do you find Skelthsea so far?'

'I like it very much and I imagine it's even more beautiful in summer.'

'It is, but coming from the mainland we must seem very rural. I expect you had a busy life back in Edinburgh.'

I sipped my tea. 'Not really. We lived a few miles from town and I spent much time at home with my sister.'

Mrs Argylle's face fell into an expression of sympathy and I suspected that she already knew some of my history. 'How brave of you to come all this way to somewhere so different from what you know.'

I did not tell her that desperation had brought me here. That I was not brave at all but had fled from the shame and grief I felt, and for one terrible moment, I saw the flames that had leapt from the windows. Had I screamed? I seemed to remember my ears ringing, or was that the fear of what I already knew I would find? Something of what I felt must have showed and she reached out a hand and placed it over mine.

'I can see that you have endured much. It is in your eyes.'

Her gesture brought an ache to my throat.

'You must be gentle with yourself, Miss Swansome. The past is sometimes better left. Now tell me, how do you find Iskar?' There was more than an idle curiosity there.

'Miss Gillies is very welcoming. Everyone is.' I paused and she raised her brows a fraction. 'But I do not please Greer somehow.'

'Greer.' She leaned a little towards me, a serious look about her. 'Greer is a complicated soul.'

'How so?' I asked.

Mrs Argylle shrugged. 'I would call her troubled. Are you telling me that she is not fully polite to you? Did you know that after Hettie left, it was mainly Greer who had charge of the twins?'

I shook my head and tried to imagine Greer with her sullen manner taking appropriate care with a child such as Mary. The thought made me sad.

'I found it a surprising choice, but in many ways it made sense. Greer lived at Iskar and must have got to know them well after their mother's death.'

It explained in some way Greer's animosity. Had she nursed the hope that she could replace Hettie? How unwelcome my arrival must have been, returning her to her more lowly duties.

Mrs Argylle did not speak for a few seconds. 'But if I may give you a little advice, Miss Swansome, I suggest you take care with Greer. She is not one to be crossed.'

But I believe I knew this already.

'And how do you find Mary? She is, after all, the reason you are here.'

'I like her, very much – and I'm confident that with the right attention she will get better,' I said.

'That is what everyone hopes,' but there was a note of doubt in her voice. 'I see you have a strong character and a good one. Miss Gillies has chosen well.'

'What happened to Mary's mother?'

'Too sad, too sad. Evangeline took a bad fever although she had been in ill-health for some time.' A shimmer of grief sparked in her eye. 'She always had a weak chest. It was not an easy death.'

'And her father?'

'A wound took an infection. The children were only three or so. Mary has indeed been through so much.'

Sensing her discomfort, I glanced about the room. 'You have made a lovely home,' I said, and as Mrs Argylle let her own gaze roam, I took the opportunity to study her. She was neat and strong with translucent skin and pale brows and lashes. A bump halfway down her nose made me wonder if she had met an accident as a child. Her manner was lively and keen to please and there was also a little something of mischief and humour. With pleasure, I realized that she was some-one I could grow fond of and hoped that we might become friends.

She leaned forward, polite curiosity and something a little keener shifting in her eyes. 'Has Mary attempted to speak since you came?'

'She has not spoken yet,' I said and placed my cup down, 'but she has laughed and there is a compan-ionship between us.' There was pride in my voice. It was good to share these things, since with Miss Gillies conversation about Mary was generally met with a change of subject.

'This is wonderful, Miss Swansome. Now that you mention it, I believe she has an easier look about her.'

'Thank you,' I said with warmth. I paused, recalling the conversation I had overheard that first morning and how Miss Gillies was so reluctant to have her nephew talked of. 'I have wondered about William. Miss Gillies does not like him discussed but I am curious as to how he met his accident.'

'They have not told you? It does not entirely surprise me – Miss Gillies is very private – but it's no secret. William fell from Stack Mor. That's the cliff you see to the east. The highest on the island.'

I recalled the peaks as viewed from the boat. This was not what I had expected. She must have seen the shock on my face. It was too easy to imagine the impact of rock.

'What was he doing up there?'

'William was a great one for wildlife. He made collections – feathers, butterflies. I'm sure you know the sort of thing. Climbing down from Stack Mor is an island tradition here. Eggs are taken to be pickled for the winter. In summer, the men take the young fulmar for oil and food.' Her eyes would not meet mine.

'He was allowed to do such a thing? Was there nobody watching him?'

'Hettie had left, and although Miss Gillies charged Greer with their care, Greer also had duties that she sometimes had to undertake at the house and the children were often about unaccompanied. William fell either at night or early morning.'

'I assume Mary was not there.'

'No, indeed not.'

I did not say anything, but that sliver of dislike for Hettie pressed at me harder. 'If Hettie had not left as she did, William would still be alive.'

'He would,' she acknowledged.

'And the twins must still have been grieving terribly for their mother when she made that decision.'

'It was selfish. One of the big American ships stopped here for trade. She went down with the children and showed much interest. I believe she was found flirting with one of the passengers. The next morning she had packed her trunk and gone. She left a brief letter but did not say goodbye, even to William or Mary. Such thoughtless cruelty.'

'How long had she minded them?'

'About four years, I should say.'

'Why would she leave? Miss Gillies said that she took her mistress's death very hard. Was that the cause?'

'She did, that is true. Who knows? She was young and if she was ambitious there is very little here at Skelthsea to help her on that path. I imagine America would have been an exciting prospect for a woman such as she. Perhaps that attracted her. She was pretty too. I imagine she wanted a husband and the men here may not have offered enough for her taste.'

'Had she made no alliances on Skelthsea?'

'I did not say that. She made several conquests during the years she was here but, as I said, I believe she was ambitious.'

'Do you have children?' I asked.

She turned half away from me. 'That was never to be.'

Embarrassed, I took up my cup. 'What was Mary like before? Before William's accident?'

'I was great friends with their mother. She was my closest friend here.' She took an album from a table and placed it on my lap. 'Happier days, alas, but you can see the twins as they once were.'

It opened to a photograph of the Argylles, but further on, I reached a print of a woman and Mrs Argylle seated on a rug, half shaded by the circles of their parasols.

'That is Evangeline.'

Her likeness to Miss Gillies was evident immediately although she looked happier – an easier character than her sister, I suspected. Without a scar to mar her looks, she was clearly very lovely.

'Evangeline was always the most beautiful, even before Miss Gillies' accident.' Her gaze lingered on her friend and although only in black and white, the colour of their happiness lifted from the paper and into the room. I could see its cheerful memory reflected on Mrs Argylle's face. In the background, wind bent the sea grass where William and Mary were crouched over patch of sand in matching sailor suits.

The next was of the twins looking to be seven or so. Seeing Mary gave me a pang; the girl here was not the one I knew. This was a child with an open, interested expression and joy on her lips, unrecognizable from my charge.

'That is their house, Gulls Cry,' Mrs Argylle said, pointing to the cottage behind. 'They lived there until Evangeline died and then moved to Iskar. It's the one along the far end of the sands.'

I looked at it with interest. Mary and I passed it most days. Further back at the entrance, a figure stood on the path with a jug. I did not need colour to know that her hair was red. It was Hettie. She wore a top with a round lace collar and her hair was so loosely held that tendrils fell about her cheeks. The camera caught her looking up and I was able to meet her eyes, black in her white face. It seemed to me that her gaze passed through the paper and fixed me in its sights. She did not smile but her lips were full and well shaped, her figure good. She was prettier here even than in the photograph I had been sent by Miss Gillies. At her throat was a large cameo; one of her fingers touched the rim as if advertising her prize.

Mrs Argylle laughed. 'How she loved her brooch.

It was a fine piece, an heirloom,' but her expression was playful and I wondered if she believed that Hettie had acquired it in another manner.

I thought of Hettie, the way her confidence showed in that direct look she paid the lens. She was not like me, I thought. She seemed like one for whom life had been easier and she had not had to question much.

Light was beginning to fail and the room was seeping into darkness. The clock on the mantelpiece chimed.

'Goodness, is that the time?' she said.

I rose, reluctant to leave. As if reading me, she took my hand warmly.

'I like you, Miss Swansome,' she said. 'I am pleased that you have come to live amongst us.'

The room had grown more chilled and outside the horizon was shrinking. Streaks of red scratched the sky. I made my way down the path. Below me, fish were being brought in and a group of women worked at untangling nets. The sounds of life filtered into the late afternoon, a song, a laugh, the squawks and chuckles of chickens from the coops, and I slowed my step, taking warmth from the proximity of so much industry.

On the beach it was silent, the sea almost still. The smell of Iskar's chimneys sharpened in the air. Turning,

Stack Mor and its soaring angles cut into the cloud. Had he called out as he fell? I imagined the rush of wind against his skin, blinding him to sound, the granite rising to meet him and the final desperate thought that hung in his head.

7

As I neared Iskar, my eyes drifted to Hettie's window, and something against the pane stopped me in my tracks. I pinned my attention to it and was assailed with a spike of disquiet. As I tried to mould what I had witnessed into cloud, whatever it was moved closer to the glass and revealed itself. There in Hettie's window was Mary herself, her expression solemn as she gazed back down upon me. My heart stilled in relief. Then, as my vision adjusted, another shape appeared behind her, before the light failed and cast the image into darkness.

Once in Iskar's hall I waited and listened – there were no voices and I took swiftly to the stairs. Faintly, the wind fluted far off, but somehow the quiet inside felt unnatural, as if it were the pause in speech before a piece of bad news was to be imparted.

Greer was standing at the foot of the staircase leading to the old nurseries. Pausing out of sight, I waited for her to move on, but she remained quite still. Too still – and there was something in her stance I did not like. I could have trodden more heavily, announcing my presence, but somehow I found myself creeping backwards to watch her through the banister rails.

It seemed too long a time before she moved, dragged a hand down her apron and walked to the servants' staircase. Was it Greer I had seen behind Mary in the window? And was Mary still up there?

With Greer gone, I climbed to Hettie's room.

It was empty. As I went to leave, my eye caught on something on the dressing table. I lifted it curiously. It was a wooden doll, only inches in height, as might belong to a doll's house. It was faintly warm, as though Mary had left it only a moment before. I turned the doll over and noticed with a grimace that its face had been scored away, leaving it entirely featureless.

I shuddered. Imagined Mary up here by herself with this strange object. Only she had not been alone.

I went to the playroom where I perused the shelves of books, dolls and wind-up toys. Light was fading and shadows stretched across the shingle. With each item I picked up, I felt the negative space of William, as if for every object that had belonged to him and been removed, the air was occupied with only its absence. I recalled the conversations I had begun that had not brought more knowledge of him. Yet I could not imagine what mystery might be contained in such reticence.

My elder brother, lost now in the distance of time, I remembered only as a ringing voice and pair of clumsy boots that swaggered through the rooms of Swan House before a fever took him at seven years. At times it did not seem possible that they had all gone, leaving only myself. I recalled the nights at the MacAllisters' home in Circus Gardens after Clara had died, the sound of the blackbird on the ivy, how each breath cut my heart like a knife.

The picture I had drawn that morning of Mary lay on her table. It was instantly clear that it had been altered. Behind the portrait of herself, Mary had added two additional figures – a woman with a thin face and a boy. They stood just at her back as if they could have reached out a hand to touch her shoulder. I knew who they were – who they must be – by their dress. As before, they bore no faces. In my mind, I sketched them in: Willam with his soft cheek, and Hettie, her eyes dark and thickly lashed. Again, something distasteful plucked at my belly. I placed it down and then, still dissatisfied, turned it over so that I did not have to look at it again.

When I left the playroom, Greer was coming up the hall again to light the lamps. I stiffened and uttered a pleasantry, but in spite of my efforts she passed me by in tight silence.

The wicks in the lamps hissed and glowed from behind the glasses and in determination, I followed until she reached my room where she began to clean the grate. Sitting at the dressing table, I pretended to study my reflection although it was her that I watched. I began to unpin my hair which was already half out of its moorings.

'It's cold out again,' I said.

She gave a nod and stood. Her lids were heavy with discontent.

'It's strange.' I laid the pins upon the table where the thin sound of them punctuated the air. 'It's my experience that people remember those they loved by keeping their things within easy remembrance.'

She picked up her bucket. 'It's not for me to say.'

'What was William like?'

An expression I could not read slid over her features and for the first time, she looked a little afraid.

'He was . . . He was very like Mary,' she uttered finally.

'Were they good friends? He and Mary?'

She shrugged. I felt a tug of exasperation. 'And Hettie? What of her character?' I pressed.

She scowled, moved to rest her weight on the other foot. 'She was just Hettie.'

'She was very pretty.'

Greer flushed, 'If you like that sort of thing.' She moved to the door.

'And was she good to the children? Leaving them when she did was unkind.'

She tipped her chin higher and her eyes narrowed a fraction. 'That's mainlanders for you,' she said, and leaving the barb in my skin, she left.

That night, as with most, I became aware of Greer's step upon the corridor. My eyes had begun to droop but the anticipation of her melody brought me fully awake. I found myself studying the rhythm of her pace, the slow swell of the tune and how as she drew near it seemed to pick itself out of the air and hang in the silence as if it were the only thing to inhabit it. Her step paused outside the room and her voice fell to quietness. My breath rasped in my ears and my neck prickled. I turned on the pillow so that I faced the door even though the room was fully dark. I waited. I waited longer; time did not seem to obey the usual rules. And then a noise came from the attic – a dragging and rattling.

I could not identify it and yet it was somehow familiar. It tapped at some memory I could not place. Then the sound changed as whatever moved met a thinner surface with an echoing space beneath. And somewhere, far off, a whistle hissed over the sound of waves.

8

The days began to run into each other and although there were things that caused me uneasiness, much brought pleasure too. I went often to Mrs Argylle who never failed to welcome me with warmth and renewed interest. Gradually, the faces of the other islanders became familar and I was able to put names to them. It was one evening, after church, that the woman who had paused to stare at me during that first service fell in beside me on the way back to Iskar.

'I'm Ailsa,' she said by way of introduction. 'You are Mary's new nanny.'

'I am.'

She nodded towards Iskar, 'And how do you find your job?'

Mary was ahead with Miss Gillies. 'I'm liking it,' I said. At that moment the clouds drew together and it began to rain heavily. People parted company and increased their pace towards home, pulling up hoods and scarves. Ailsa raised her collar, wiping drops from her face.

I thought she had gone when she caught my sleeve

and leaned towards me. 'I live in the cottage off near that sheep fold.' She pointed beyond the main street to one set apart. 'Come visit me.'

I thanked her and moved off, but she held my cuff firmly and her unseemly grip on my coat and the familiar tone of her voice filled me with discomfort. I tried to pull away, mentioning the weather and the need to be home, but she brought her lips so close to my ear that I could smell her breath.

'All is not well at Iskar,' she whispered. 'You feel it, don't you?'

I opened my mouth but could find no answer and instead hurried to the beach, reaching the path a little breathless. I paused to refill my lungs and shivered. Stupid, I told myself, stupid, but against my will Ailsa's words and the gravity of the way in which she spoke them slithered inside me and set up an echo to my own half-considered sentiments.

Later, after I had put Mary to bed, I made my way to the drawing room and found Miss Gillies sipping something hot and herbal in a tall glass. 'Forgive me, Elspeth, I have a headache tonight.'

'Is there anything I can do? Would you like me to read to you?'

She shook her head. 'I will just sit here for a while and let this magic do its work. Mrs Lenister is a marvel with herbs.'

I settled myself on the sofa and took up some embroidery.

Miss Gillies lay back against the cushions. 'I should probably have told you earlier, but a doctor is coming tomorrow to see Mary.'

'A doctor?' I asked in alarm.

'He's a friend of the family and I consulted him about Mary after William's death and how it had affected her.'

She wiped a stray hair from her brow. 'We arranged that he would come within two months to assess her improvements.'

'And has she made them?'

She did not answer.

'What was the doctor's diagnosis?' I asked.

'Mary is not quite herself in her head, is she, Elspeth? Not quite as she should be.'

I understood her meaning then. He was not a doctor of the body but of the mind. This idea did not sit well with me.

'Mary is a child who has experienced too much death, that is all,' I said.

'Of course, but one would have hoped that she might at least speak.' Her eyes were slightly accusatory.

'Love and kindness will heal her in time, surely.'

'And that is where you do so well, Elspeth.'

I thought about that answer and how she appeared

to have passed all responsibility for that role to a stranger such as myself. Although she cared for Mary enough to undertake lessons, there was too often a tone or a look that she passed to her niece that showed not love but dislike.

She rang the bell for Greer to attend and rose carefully from the chair, resting a hand on the arm to steady herself. 'Excuse me. I'm a poor conversationalist this evening and now I must be off to my bed.'

Greer came so quietly that I did not hear her until she was behind me and there was a murmur of conversation followed by the sound of retreating feet.

When the clock chimed the half-hour, I decided that I too would retire. The corridors were silent. At Mary's door, I paused and put an ear to the wood. From somewhere inside came whispering. Carefully, I went in. Mary lay on her back, one arm flung out of the covers. She was clearly asleep. I scanned the room – nothing. Apart from the faint rush of the sea, there was only silence. Yet there remained on my ear the memory of a childish tongue and the undulation of speech. Mary's face was still, so still it was hard to believe that she had uttered a word as she dreamed but I stayed listening until, growing cold, I left for my own room.

In my bedchamber, I was aware again of a creeping of some other feeling – beyond numbness and pain, beyond Greer's dislike, not anything I could name, but which was uncomfortable. I tried to dismiss it and told myself that I was being fanciful, but the spaces bloomed unhealthily.

I emptied water into the basin and cupped it in my hands, bringing it to my cheeks. But as I enjoyed the iciness of it, there was a deeper chill at my neck, something colder than the room itself. Letting the water trickle through my fingers, I gazed down at my reflection and in the ripples another face loomed over mine. I gasped, shock bolting through me. I blinked, willing my mind to dismantle the image, but when I opened my eyes again the face shivered into something pale and smooth – no features, just an oval globe. I whipped round but there was nothing, just the thudding of my heart and the bitter air.

With trembling legs, I got beneath the covers. The rasp of fear was in my quickened breath.

'All is not well. You feel it, don't you?'

9

I woke to the dripping of rain and a puddle of water that had gathered on the sill. When I opened the curtains, the island was smudged by a steady downpour. Angus passed below with sacks, shoring up the gaps beneath doors. The boat was due and with it the doctor.

Sometime overnight, the position of my shoes had been altered. I distinctly remembered having left them by the fire, sole to the floor. Now they were leaning against the wall – toes to the ground and heels to the wallpaper. Had Greer done this when she came in early to leave water? I imagined her, in my room, moving them for some purpose known only to herself. There was nobody else I could think of who would undertake such a thing. If it were Greer, had she paused and watched me sleeping? The idea of her louring presence looking down on me in silence, left me faintly sick. I finished dressing and went to raise Mary from her bed.

She was deeply asleep. There was a faint darkening of the skin beneath her eyes, a shadow that spoke of restless nights, and I frowned, recalling Miss Gillies' tone of the night before. A doctor. I did not like the direction my thoughts took at such a consultation. I remembered the doctor in Edinburgh's grave voice outside my door and knew what it was he suggested to the MacAllisters. Knew the place he recommended I should be sent if I did not improve. Could the same fate be in Miss Gillies' mind?

That morning, I spent longer on Mary's appearance, curling her hair and setting it with ribbons and clips and cleaning her face and nails until they shone.

Downstairs, the house vibrated with a new energy. Greer was waxing the floor and Magda polishing the fireplace.

'Good morning,' I said. I opened my mouth to mention the shoes but almost as quickly changed my mind. If she sought to frighten me then I would only give her satisfaction by letting her know that her actions had been noted.

Magda looked up, brushing the back of her hand across her brow. The smell of beeswax and lemon oil hung in the corridor and a warm scent of baking drifted from the kitchen. Miss Gillies was in the dining room, her hair neatly made into a knot at the back of her head and held with a diamond clip.

Mary seemed anxious about the forthcoming visit and I was keen to fill her morning with something distracting. The rain had been replaced with a watery sun and we left the house, taking a kite with us. At Gulls Cry I glanced up at the streaked windows but Mary kept her head down, unwinding the string on the bobbin with fixed concentration.

'Come on, Mary,' I said. 'Let's make it fly.'

The string was unspooled but the kite bounced to the sand and she gave me a nervous look.

'Hold it,' I said, 'and catch me.' And we were running along the strand like two children; I could feel the laugh in my throat at her shocked expression. She tugged at her toy and tried to catch the wind in the kite's sails, but it dipped and rose with a whim.

'It's like a mad thing,' I shouted, and she made a sound like a chuckle.

She took off again and I ran beside her, and finally, it caught an updraught and soared into the sky where a parting in the clouds revealed honeyed rays of sunlight.

'Keep it up.' And I helped her with the twine, our fingers touching as we lifted our heads to where it flew above us, brushing the heavens like a sign. No words were needed, but between us the joy of success and of the glittering ocean was shared.

As we returned, she kept her distance, hopping towards the shore then drawing back as the tide grew close. Now and then she looked at me and then she began to run back and forth along the bay, stopping to examine the ground, and as I watched, I saw that her lips moved as if in speech. I paused and listened but the noise of the waves drowned all else. I wondered what it was she said. Did she practise the sound of her tongue or was she lost in some narrative inside her imagination as children were so often wont to do?

Eventually she caught up to walk beside me. The hem of her skirt was damp and her boots showed a white line where she had been caught by the sea. I tried to read her face but she had her head down and all that I could see was the bobbing of the yellow tassel on her hat.

A wave came up nearly to our feet and I jumped to avoid it, but for a moment, beside my own reflection, there were not one but two small shadows upon the water. I blinked, looked again, but the wave had retreated leaving only the skin of wet sand. Surely a trick of the light? Yet there remained upon my memory the conviction that the shadow figures had not been the same and that one had been thinner and taller.

The ship was docking and Mary and I waited with the others to watch the passengers disembark. Reid Paterson leaned against the sea wall and gave me a slow smile. Beside him, his wife hunched her shoulders into a shawl and kept her head down.

For a while, the doctor was lost in the throng. When he finally came through the crowd, he was easy to identify with his black coat and top hat – a leather case in one hand and a silver-topped cane in the other. Angus was there to meet him. Mary's face showed no reaction. Once again, the lid had been pulled over her inner feelings. I wished that he had been able to witness us earlier, chasing up the beach with the kite, and heard her laugh. I squeezed her hand as he neared us. At first, he appeared not to have recognized Mary but then he slowed.

'Well now, Mary. This is a pleasing sight, that I should be met so keenly.'

She drew closer to me.

He gave me no acknowledgement and in the face of Mary's lack of response, he touched his hat and continued up to Iskar.

Miss Gillies wore a dress that I had not seen before of green bombazine with sprigs of embroidered mimosa. Every fire had been lit in an attempt to raise the temperature. She and the doctor took wine together in the drawing room and Greer brought tea up to the playroom where I tried to chat lightly and engage Mary in a game, but I soon fell silent.

After a while I stepped into the corridor where Greer's slow footfall signalled that she was ready to take Mary down. To my chagrin, the doctor did not request to see me.

Lunch was served in the dining room and I was

pleased that we were joined by the Argylles. The mood was cheerful, conversation lively, and even Miss Gillies found places to laugh. It was impossible to judge how the consultation had proceeded. After that, I took Mary to the playroom and it was not till later that we heard the boat's engine and knew he was gone.

Rain started again and the dripping of gutters filled the corridors with melancholia. Outside, the sky was a stubborn grey and the rooms seemed to swell with damp; I began to notice new areas of decay, patches in the ceiling that grew drops of water, and a smell of must and mould flowered in the air. It seemed a long time ago since Mary and I had flown the kite.

In spite of the extra fires, the temperature was chilled, and when lessons began and the weather improved, I walked up to Mrs Argylle's where she met me with an eager smile.

She was in an exuberant mood, showing me a pretty thread that she had received in a peacock blue, pressing a further sample into my hand and insisting I have it.

'But you look worried, Miss Swansome. Is there something the matter?'

'If I am honest, the doctor's visit has made me anxious.'

'Why is that?' she asked.

‘I cannot ask you to break any confidence you have with Miss Gillies, but I am concerned about what sort of a diagnosis such a doctor might make.’

‘Ah,’ Mrs Argylle laid down her thread, ‘I think I understand where your mind is going, Miss Swansome. You worry at the cure?’

I nodded.

‘Try not to be too anxious. Miss Gillies is only doing what she feels is right. It’s her duty to see if something can be done to help her niece. She is pleased with you, I know that much. There is still a while before he returns.’

‘Returns?’

Mrs Argylle flushed. ‘I’m sure Miss Gillies will tell you, but yes, he will be coming again in three months if Mary does not make the necessary improvements. Has she shown further signs of speaking?’

I had to shake my head.

That evening, as we made our way to chapel, the mood amongst the villagers was buoyant and we stopped more than once in conversation. The boat had brought letters and money and there was an air of celebration.

By the time we arrived home, everyone was soaking: Miss Gillies shook her hat, freckling the tiles with droplets. I unbuttoned Mary’s coat to find her frock sopping beneath. Bobbity had been kept dry beneath the bodice of her dress.

In Mary's room, I helped her out of her wet clothes.

'Are you worried about the doctor?' I asked.

She paused and her fingers found my hand which was busy at her buttons and stilled it. Her eyes sought mine and for a moment, it seemed as if she were trying to tell me something and her sadness was like a sudden piece of glass to my heart.

When she was changed, I told her to join her aunt by the fire and, sighing, I tucked her clothes under my arm and went to my own chamber. As I placed Mary's things on a stool, something rattled in one of her pockets. Assuming it was shells, I felt around, but what I drew out was a pebble. It was dark and flat, but on its surface a figure had been drawn and on the other side a symbol had been scratched. The whole stone was wrapped around with hair. I could not imagine for what purpose such an object had been created.

But as I studied it, my discomposure grew. I opened the window and threw the pebble as far as I could.

My eyes skated up to Stack Mor and I noticed, for the first time, a strange jutting circle of rocks that lay on the ground next to it, too regular to be manufactured by nature. I moved closer to the glass but it was nearly fully dark and the rain began again, obliterating the view.

In the hall I stopped Mrs Lenister.

'I noticed just now some stones up by Stack Mor.'

She gave me a curious glance. 'That's the Fiaclach. It's been there for hundreds of years, they say.'

'What's it for?'

'For? Well now, it's famed in its own way. We had a group of Edinburgh scientists come once. It was made for pagan uses, you know, all that sacrifice and old religion.'

'I had not known that.'

'If you go, take care. It gets rare windy up there and the cliffs are dangerous, so take caution if you want to explore. Always take a stick, Miss Swansome. I can see you like your walking, but the mist comes down awful quick and the ground is rough. You wouldn't be the first to take a nasty tumble.'

Coffee was served in the drawing room and I took my place on the sofa and studied the dark drapes and hangings with distrust. Mary had gone upstairs in search of a game. Miss Gillies chattered inconsequentially and all the time, unease settled into the very pores of my skin. I thought of the pebble and wondered if this was some game Mary undertook with her brother.

'Did Mary and William play together much?' I asked.

'Well, yes, they did by necessity, although Mary

preferred the house and her toys and William the island.'

'Did they have shared pursuits? An interest in any of the same toys?'

Miss Gillies looked up from her work and pierced me with a stare. 'Why do you ask?'

'For no reason,' I lied, 'I was merely interested to understand more about Mary and her grief.'

'There is nothing there for you to learn,' but her hand worried at her brooch and I felt the vibration of secrets.

Mary returned and we played for a while and later I put her to bed. Leaning down, I gave Bobbity a kiss. She watched me intently and lifted the doll and touched its lips to my cheek. She smiled then. It was the tiniest movement and I knew that her face hid more than it showed, but its power dissolved some of my anxiety. I was gripped by a fierce tenderness for her that swelled in my throat and I had to turn away to hide the volley of emotion.

'Goodnight, Mary,' I said.

The passageway was chilled and a thin wind rattled at the panes. Placing the candle in my own room, I went to stand at the window. With only the faintest of moons there was little to define the island – just a line here or there, marked in moonbeam. My gaze wandered up to Stack Mor and below, in the circle of stones, a lantern lit the shape of someone kneeling. I thought of what I had so recently learned and pressed my face close to the glass where the chill floated on my skin.

As I stared, I had an instinct that my scrutiny was returned. I stepped quickly back into the room, where I sat, my heart pattering as though I had been caught out in some misdemeanour. When I stood and looked again, there was no light, and the figure was gone.

10

The next day brought more rain and after a morning in pursuit of a jigsaw that was clearly uninteresting to both myself and Mary, I suggested a game of hide-and-seek. She brightened a little and went off first to hide. I sat on one of the chairs in the great hall and tried to trace the direction of her footsteps, having instructed her not to use either her aunt's or the servants' quarters.

I closed my eyes and duly counted to one hundred and then, calling as I went, climbed to the first landing and began to explore. As I opened each door, I announced myself with a, 'Now I wonder if someone might be hiding here?' And as I crossed the rooms, walking deliberately heavily, I whipped away valances, yanked curtains and opened cupboards with a flourish, sending dust flying into the air.

I was struck again with how little remained to demonstrate that William had once lived here. After Papa had died, his pictures stayed in their frames and the rooms my lost family had inhabited were left to grow dusty but unchanged after their passing. To sit in

Papa's chamber and to gaze at the familiar objects – to run a hand along a swathe of tweed or open the bottle of cologne and remember his scent – was a profound solace. I could conjure my father's voice from his ivory shaving set or the slant of his handwriting on the nature notes left on the nightstand. All ashes now.

In spite of a thorough search, Mary was nowhere to be found, leaving only the old nursery rooms and Hettie's bedroom. I paused at the bottom of the garret stairs and listened. Silence. My chest was tight as I climbed. The schoolrooms contained no place in which to conceal oneself. From the windows the island blurred in the rain and gulls circled low over the valley. The air was smoky and dense.

The only place left now was Hettie's bedroom. My feet resisted the exploration, did not want to witness again the aged comb and the faint scent that remained. In the corridor outside, I called out, 'Coming to get you.'

Then it came – the clatter of feet and the lightest of sounds that could have been the readjusting of a body in its hiding place. I thought, with relief, that at least the displeasure of revisiting where Hettie slept would signal the end of a game that had become laborious, so I threw open the door with something like triumph.

'I know you're here,' I sing-songed, but as I lifted bed linen and explored cupboards to no effect, I felt the growing of dismay. Finally, I stood by the hearth knowing, but not wanting to acknowledge, that there was nobody here. And what had I heard? A trapped pigeon in the attics? A rat? Some strange echoing of sound from another part of the house? And then I noticed, on the dressing table, something that had not been there before. I went to investigate and found a black stone, too similar to the one that I had discovered in Mary's pocket to be anything other. It could not be – surely not. I had thrown it full into the undergrowth. It was cold and heavy on my hand. I swallowed uneasily and placed it in my pocket to lock in my drawer.

I stepped to the door, keen to be away, when the whistling came whining from some place high up; it stopped and started as though manufactured by intent. It caught me and trapped me, releasing some profound dread.

I leapt down the stairs, catching my breath for a few moments at the bottom. Once I had recovered, I began to feel annoyance. Where was Mary? And then – there she was – standing in the hall beyond, that unreadable expression in her eyes, the mouth neither smiling nor displeased – a face that gave me nothing.

'And where have you been? You must have hidden

well.' Although I attempted to rein in my irritation, there was exasperation in my tone. But of course she did not answer and I felt the further stir of emotion, already unsettled by what had just happened.

'Where on earth were you? I looked everywhere. You didn't go into the servants' or your aunt's quarters, did you?' She shook her head and I smothered a sigh, wondering with anticipation if it might soon be time for lunch. I followed her along the passageway where she turned into her bedroom.

'But I looked here,' I said.

She pulled me to where a tapestry hung against the wall, drawing it aside to reveal a door. With a gesture, she indicated that I open it. It swung wide to a small room with walls of stone. The only light came from the dull beam that fell through a thin aperture, about which webs hung like charred lace. I shuddered. Had she hidden here? In the dark and alone?

I did not want to stay. I knew what this was – a priest hole. Swan House had had one too, but there was something about this particular space that made me feel faintly sick. Even as a child, when I had loved to hide, to have sought a location such as this would not have appealed to me. It was unnatural. I thought of all the secret places that might lie within Iskar, imagined Mary or William crouched in the smothering blackness waiting to be discovered or smuggling themselves away for the sheer stealthy joy of it. Worse, I pictured them together with a hoard of pebbles. I looked down at Mary. Her expression was untroubled. Uncomfortable as I felt, I did not want to think of her as the child who had held the chalk that inscribed the figure on a piece of stone.

I backed out, knocking my heel against something; it skittered across the floor before rolling on to its back. It wobbled and settled. One rusted wheel spun for a further second or two until that too ceased; it was a toy steam engine, its surface patched with red paint. I looked up, catching Mary's eye. Surely this had belonged to William. Had she been playing with it? I tried to find some answer in her expression – distress or grief or something to mirror what she felt – but there was nothing to see. As I left and dropped the hanging, I thought I heard a scuffle beyond and a tinny echo as if someone had knocked the engine once again.

Later, in bed, my thoughts tangled. I was a rational being. I did not believe in the supernatural and yet I was being presented with things that I could not explain – the sounds, the odd glimpses of something that should not be there. Turning on the sheets, I buried my face in the pillow. I began to doubt my own judgement. I remembered that after Clara's death what was real and what was not became muddled and I had no longer been able to fully trust my senses. Was that what was happening now? At times, I had heard Clara's voice or felt the heat of fire on my neck only to realize that it was a false construction of grief and guilt. The world twisted out of shape and into something that no longer made sense. Hadn't I woken to the scent of smoke and the sound of my father's pipe tapping against the table? Those things had felt so tangible, more tangible than the polished floor and the daisies on the surface of the MacAllisters' washstand.

If Papa were here now, he would have answered my doubts with a laugh. I knew where he stood on things supernatural. Had we not had conversations long into the night, turning superstition, religion and philosophy on to their backs to enquire inside? The Earth was made of mineral and flora; animal life was the only sentient presence upon its surface. I would have welcomed one more glimpse of Clara, and yet that never came, except in my imagination. There was no door between life and death.

I do not know what time it was, but I woke to screaming – high and thin like an animal. Mary. My limbs were uncoordinated as I ordered them to the floor. Lighting a candle, I stumbled into the corridor.

I had almost begun to hope that her nightmares had stopped since my arrival.

She was upright on the mattress, her spine as straight as one of the hall chairs, eyes wide and staring at some fixed point. There was an unnatural immobility to her face.

I touched her arm. 'Shush, shush, it's just a nightmare.' Her cries fell to an intermittent whimper. 'Hush now. It's just a dream.' But my heart thudded.

She turned and looked directly at me, fear and something else embedded in her eyes. She began to whisper fast and low but the words fell into each other like waves crashing to the shore and being instantly swept up by the next onslaught. It was curious to finally hear her voice, but as hard as I tried, I could not comprehend her. I listened closer but the sounds she made did not follow a language I understood, and I drew back. It was clear that she was still asleep.

But at that moment Miss Gillies was in the room, her features stretched. Greer was behind her.

'Don't worry, Elspeth,' Miss Gillies said, 'I know what to do. Go back to your bed.' There was alcohol and clove on her breath.

'I can help,' I said. 'She's talking in her sleep but she's not making sense.'

'It's nonsense she utters. Just go.' There was a note of command in her tone.

'But I'm sure I should help her,' I said, 'I'm happy to.' Greer stood and watched, her arms folded and a gleam of pleasure in her iris. Although snubbed, my heart ached for my charge, but Miss Gillies was my mistress and I could not protest.

It was then that I noticed something that made my skin shrivel. On the wall opposite the bed, the hanging had been pulled to the ground and the door to the priest hole lay open again. A wash of its stale odour crept into the air. Once more, the space gaped, tunnelling to a darkness so opaque it could contain anything that the eye would not see. Horror crept along my neck. Miss Gillies turned and saw what I saw. She drew herself up. 'Greer,' she called, 'deal with that.'

Miss Gillies gave me the briefest of glances that might have been an apology as Greer swept towards us, her candle flickering. Light fell briefly into the uncovered room. And for one terrible moment, something glinted back at me from the furthest wall – two surfaces with a gleam upon them as light might make on the film of an eye.

11

Morning came and Mary was still asleep. I found Miss Gillies drinking coffee in the dining room. I took a seat opposite. Miss Gillies' expression betrayed no sign of the night's drama.

The timbre of Mary's voice still echoed in my ear, with that strange concoction of syllables. 'Does she often speak with no sense when she is in sleep?'

She looked up. 'I suppose,' she said.

'Did she do that at any other time?'

'I am not sure. It is not uncommon for the young to manufacture their own secret language. Pay it no heed.'

'And the priest hole? Are there many at Iskar?'

'There are a few. Most now boarded up for safety.'

'But not the one in Mary's room?'

'There is nothing in that one that could cause harm. In fact, I believe the children used it sometimes for their games.'

I shivered and she laughed wryly before continuing. 'Indeed, it is a little ghoulish, but you know boys – they have a thirst for these dark places and dens, don't they?'

'So not Mary?' I pressed.

She shrugged. 'In truth, I do not know.'

Mary was quiet after breakfast. No mention was made of the nightmare and we went down to the bay, walking to where shallow caves lay. I felt that I was peeling back the layers of her distress and discovering new wounds. The sea hissed and my thoughts wound around themselves in endless circles. There had been something about her terror in the dream that seemed to touch at something even deeper, and as I mused it came to me suddenly and I turned to regard her.

Fear, I thought. She was not only unhappy; she was afraid. Afraid of what? Of death? Of Miss Gillies, or was it Iskar itself? And I did not like the way that idea numbed on my lips.

On the way back, we went to the shop – although facing Paterson was a duty I would have gladly avoided – and I bought some chocolates. I shared half but the rest I kept, conscious of the urgency that she must speak. I put them in my pocket with the idea that I might use them to try and coax Mary from her silence.

That evening as she lay in bed, I reflected on the guilt that was ever present in me for Clara's death – not, in truth, that it had been my fault. I took Mary's hand. 'Sometimes we blame ourselves when someone we love very much dies, and I know how you must have loved William.'

Her eyes grew rounder.

'I had a sister too, you know. You remind me of her a little.'

She stared up at me with wonder and I pulled the locket from my neck and opened the lid. 'Here, see. This is Clara.' My voice faltered.

Mary reached out a finger and touched the surface of the casing.

'She was younger than me and I looked after her.' I read the question in her eyes. 'She died. Earlier this year. See. I know what it's like.'

After gazing at it for some time, she leaned briefly against me and a rush of gratitude sent me across the corridor to my dresser, where I found a bangle, before crossing back to Mary's room.

'This was Clara's,' I said, and she regarded it with awe as I slipped it over her wrist.

Her face tightened and I thought that I had made her angry, but then two tears began to make their way down her cheek.

'Oh Mary,' I said, pulling her to me. I had never before held her so close; her ribs rose and fell beneath my palm and something in me seemed to break open a little, somewhere I had kept locked since Clara's death. After she had stopped, I continued to hold her and a conviction came to me that I had always been destined to come to Skelthsea and to take charge of Mary. That somehow, we had been meant to find each other. And we sat together in the chill of the room, sadness splitting the air about us.

'You can't beat weather like this to work off the cobwebs.' It was a new day. Outside, a brisk breeze set clouds scudding across the sky, and Mrs Lenister had come to offer tea. Mary had gone to lessons and I decided that she was right. I needed a walk to clear my thoughts. I left Mrs Lenister and gathered my coat and boots. In the village, clothes and bed linen swung on lines at the back of the houses and a group of women were banging a blanket against a board to thicken the fabric.

I strode up the valley and headed left towards Stack Mor. Ever since learning of how William had died, I had a curiosity to see where he had fallen. As I walked, the landscape grew wilder, the land cropped close with gorse and boulders. Occasionally, I came across the crumbling remains of a sheep fold and patches of bog. Soon, the activity of the village grew distant and the air was filled with the scream of gulls and the fiercer, icier pull from the sea. As I climbed, it grew overcast. I was unsure as to whether to go on or return, but in those few moments of contemplation the first drizzle began to fall.

Disappointed, I made my way back along the ridge towards the main path. Before reaching the church, I passed the graveyard and curiosity made me open the gate and enter. I walked slowly between the stones, reading names where they existed, but so many were difficult to decipher, being overgrown with moss. Finally, I came to Evangeline's in a square of flattened sea grass. She lay beside her husband beneath a marble ledger. The roses that had been etched around her name were gathering lichen spores and sand. 'Beloved wife and mother to William and Mary.' Beneath that were faded names I could not read, and next to them lay several older, smaller graves showing the infant deaths of other children. I was aware of a silence that fell deeper than the lack of sound – a primordial hush.

Iskar was in my eyeline, its gabled roofs and casement windows cast to grey. Mary would be somewhere inside, lost in the weave of corridors, and suddenly, for no reason I could fathom, I was afraid for her. The wind was chillier then. The land beneath my boots, frozen hard above cold bones.

Clouds drew together, sending out another spit of rain, and I shivered. Huddling deeper into my coat, I came to William's plot but I barely read the stone; it was what lay below it that drew my attention. On the patch of soil, beside a posy of long-dead flowers, lay the remains of a hooded crow, strung from the ornamental angel on a piece of twine. For a moment I had to convince myself that what I saw was real. It had clearly been there for some time – reduced to a macabre arrangement of bone and feather. This was no unfortunate animal come to grief by accident. This was an act of deliberate insult. Shock and disgust dripped into me and I recalled Ailsa's words: '*All is not well at Iskar.*' Iskar, where there remained the strange absence of anything that might remind his family that he once existed. Everything that I had learned gathered into a knot.

I had sensed from the first that the reaction to his death was unnatural and I realized that this act of desecration must have been accepted by the whole island to remain, as it was, in plain sight. It came to me then – William was not beloved. Miss Gillies did not miss him and perhaps Mary neither. He had been disliked, disliked with such force that even his grave did not deserve the respect reserved for those who had passed. What could he have done to earn such loathing? It was only then that I noticed it. I crouched low, pushing at the object half buried by earth. But there was no mistake; it was a pebble such as the one in Mary's pocket.

Shock left me stunned and my eyes slid back to Iskar; mist shivered above its roofs, and shadows crawled into the recesses.

The building stood so quiet, so solid, and yet it carried things inside its skin, things so dark that nobody dared speak of them – and I knew in my bones that something was amiss. It was as if I had been led up some false path.

And then it came – that thin whistle, sounding like the wind but not so earthly.

But around me there was no one, nothing. My mind went back to the times I had heard it before and nausea rolled in my belly. There was menace and intent to the sound, yet the landscape was empty of a cause. I stumbled out of the gate and back to the house, because I had nowhere else to go and because the rain had begun in earnest, leaking in through my collar and on to the clothes beneath.

12

As I entered the hall, Iskar felt different. The resonance of Mary's grief was replaced with something else – the mute unacceptability of William's life and the acceptance of his death. And Mary, where was she in all of this? What had William done to deserve such censure and was Mary implicated in some way? I had witnessed how Miss Gillies looked at her.

As I passed the morning room, the door was open a crack. Miss Gillies and Greer stood together at the window, their backs to me. Miss Gillies was leaning down and saying something into Greer's ear. There was an intimacy to their pose that did not conform to the mistress-and-servant template and something, too, that spoke of long-held affection.

I made my way up to my room where the fire was not yet lit and pulled one of the bed covers around me. My hands trembled with dismay.

Never on those nights when I had imagined Mary and the new life here had I considered that something might be so amiss, something beyond the distress of my charge, something harnessed to the very air.

I burned with loss then. I missed so wholly my father's voice and Clara's touch – the way her fine hair fell through my fingers as I dressed it. Grief returned and battered me. Perhaps, I thought, mourning could never be fully emptied.

It was later than usual when Mary went to bed. Downstairs, Miss Gillies was not in her usual place. Instead, she was standing at a far window staring into the blackness. She turned and we sat in our customary seats, although she made no attempt to lift the embroidery from her workbox.

My heart began to beat harder. 'I visited the graveyard today,' I said.

She met my gaze steadily, showing that she already knew as much.

'Yes, I was told of that. I've not been to his plot myself but I suppose that much is obvious, as it has not been tended.'

I wondered if she knew about the desecration. For a few moments she struggled with her words.

'There is no more use pretending that all was right with William.' Her expression grazed mine, looking for sympathy, but I had witnessed her lack of affection for Mary and was not moved. 'I was not brave enough at first to give you the truth I owe you. I hoped that all that was unpleasant had passed but I was naïve. You would always have learned it sooner or later. The fact is: he was not as other children.'

The anticipation of some further shock left me chilled.

'There were whispers about him. He was not a good boy.' She gave a humourless laugh. 'Why can't I just say how it was?' She looked to me as if for an answer, and I felt finally that I would hear the truth. She put a hand to her high lace collar as if she might find an explanation there.

'How was he not like other children?' I recalled his face from the photograph, the curve of his long lashes, the boyish kink to his hair.

'They were beautiful babies – but then, this past year – the rumours about him. They were terrible. He liked to be cruel.'

Cruel? And I thought of Mary. Was he cruel to her? 'What were the rumours?'

'That he had gone bad. That he was unkind, brutal even – that he had unnatural interests.' Her voice flattened with dislike, 'Things happened here on Skelthsea. Strange and horrible things.'

'*All is not well at Iskar.*' Ailsa's words slipped inside me again. 'Horrible things?'

'Things that you would not expect of a Christian child,' she said tightly. 'An interest in exploring other gods, if you understand me, Elspeth.'

Her statement was issued almost as a challenge, a dare that I might deny. I thought of the pebbles and the stone circle. 'You mean like witchcraft?'

She gave me a dark look and a darker laugh. 'As ridiculous as that sounds, yes, that is precisely what I'm saying.'

And I had to turn away to hide my mounting dismay. How could a child of nine be involved in pursuits that linked him to witchcraft? 'But he was so young. Is this proved?'

Her eyes grew hard. 'Don't you think that I, of anyone on the island, would question the veracity of this? Of course I questioned it. I'm an educated woman, not one of the hysterical servant class. I do not believe in witchcraft, naturally, but there are a few here who sadly still cling to very old beliefs in spite of the church.'

I felt chastised. 'Of course,' I said. 'It's just that I am so shocked.'

'The proof was witnessed by those I trust.'

'What proof?' I hardly dared hear the answer.

'There were animals found within the Fiaclach – gulls, mice, hare, that kind of thing – they were all slain and laid on the altar stone. They had not been killed easily.' She could hardly look at me. 'It was William who tortured them.'

I felt a punch of horror.

'And Mary?' My heart flicked uneasily.

'Not Mary. There were no rumours of her and she showed no sign that she was the same.' Her tone was measured, but in her face there was some doubt.

I realized then that I could not imagine Mary capable of such acts. Only a few days ago we had caught fish in our nets and I had watched how tenderly, how carefully, she placed them in our pail and later, how gently she released them. And the pebbles, I reasoned: if they were sinister they must have been William's and she played with them without knowledge of their implications.

'But where would he have learned such a thing?' I asked.

Her eyes slipped to the window as if she hoped for some intervention.

'I am sorry to say that it was Hettie who led him astray.'

The shock left me winded. 'What do you mean?'

Her gaze was unsteady. 'In those last months, Hettie was believed to be a witch.'

Hettie. And once again I was in Hettie's room with the unsettling sense of her presence, of being watched. The pebble on her dressing table.

'She too was seen at the Fiaclach,' she continued. 'She was seen there with William.'

I wanted to stand to catch my breath, pace the

room, to walk out into the fresh island air and cleanse myself.

I hardly dared ask. 'And before that – when your sister was alive. Were there rumours then?'

'I heard nothing before Evangeline died, but I did not see the children much.'

I looked at her questioningly.

'Unfortunately, we were not close.'

'That is sad.' I wanted to ask how such a thing came into being. For a moment I felt the sweet weight of Clara's head on my shoulder as we lay in bed, whispering about the day or our plans for the morrow.

She looked away then. 'And in truth, I did not get to know William very well when he came here. Hettie took care of the children and when she left, I had Greer. Apart from lessons the contact was not so much.'

'And when people told you what he and Hettie did?'

'I ignored it,' she said simply. 'I chose not to believe what I heard, but after Hettie had gone I found the evidence of it myself.'

I looked down to hide how much that appalled me. I wanted to ask her what she had discovered but I held back. What I had learned already was distasteful enough.

'And Greer was not able to steer him right?'

She looked up at me, something steely in her eyes, and I read the answer there.

I climbed the stairs with heavy legs, pausing for a moment outside Mary's chamber. I opened the door and looked at her sleeping face. I felt horror at Miss Gillies' negligence and the damage it had caused. Clara's bangle, which she had refused to remove at bedtime, circled her wrist. I imagined Hettie, whispering spells into her ear – slim fingers winding hair about a pebble. As I looked upon her slight form, that fear that I had felt for her at the graveyard came again pressing on my chest.

In my room, I undressed quickly and climbed beneath the covers. Above me the weight of Hettie's room drifted downwards on the air. A woman who experimented with witchcraft – I did not believe in such things, that the world could be twisted on a set of words or an object, even the death of an animal. I wanted to laugh at the idea, but the thought writhed in my stomach like a piece of paper caught in fire. I imagined her then in some room in an American city with red-cheeked children lying asleep just along the corridor. I imagined all the wickedness behind those lovely eyes and all the poison that she would let fall from her tongue, and shuddered.

I saw in my head the bird that hung from William's

headstone – its wings twisted like a macabre brooch. I pulled the covers closer and felt something hard at my neck. Reaching in, I plucked it from the pillow and lit a candle.

And there, lying on my white palm, was a new pebble, round and dark – a childish figure drawn on its surface and bound, as if in chains, by hair.

13

I was sore from lack of sleep and the day ahead settled on me like a sigh. I felt again the smoothness of the stone and the depth of dismay fell deeper still. Who had placed it in my room? Although my thoughts flew immediately to Greer and her dislike, I had no proof. And if it was her – was that because she wished to unbalance me, or was there a motive more sinister? If it was the latter, I must not let the idea unsettle me. Even if the pebbles were instruments of witchcraft, it did not follow that they had agency. I, like Miss Gillies, did not accept such things could hold real power – but did Greer believe it?

And William's falling from grace . . . had Greer played a part in that too? I noticed again that she had made a strange arrangement of my shoes and in a fit of rebellion I laid them flat on the floor with a slap.

From the dresser, I took the photograph that Miss Gillies had sent me all those months ago and studied it again. The picture was now changed by what I knew; it seemed to me that Hettie's hand, which rested on William's shoulder, had a proprietary grasp and his eyes, which had at first looked so innocent, now appeared to contain every wickedness. Hettie's expression was dark and unreadable but seemed to see beyond what we could see. I studied William and tried to imagine him undertaking the crimes of which he was accused. It was then that I noticed the flaw that threw the dark shadow across his face now bled into Mary's image. Had that been there before? I could not remember.

As I entered the corridor, my eyes were drawn to the door to William's wing. Would I find evidence there of his nature? I tried the handle but it was locked. I thought that I would feel disappointment but what I experienced was relief.

As I washed, I heard again that scrape and slur of sound above me, as if something was being pulled or pushed across the floor. My pulse quickened, and then came a rattling, as whatever it was crossed the rug and met the floorboards. I knew in that instant what it reminded me of: Clara with a toy carriage, kneeling in the hall and driving it along on the imaginary journey that played out in her head.

I took the garret stairs at a run. Hettie's door was already open.

'Mary?' My voice was swallowed in the hush. I stepped in and pressed my palm to the wall. A tinny sound came up from the floor and I looked down.

The red engine lay on its side, one wheel in a slow spin. I drew back, the breath of the room on my mouth. The corners were steeped in shadow and there was such a stillness that I knew I could only be alone.

For a second, the world seemed to turn my senses – what I heard, saw and felt did not match what was possible. The room was empty and yet, not moments before, the noise of the engine being pushed across the floorboards had filtered down to me. I backed out, scrabbling for explanations where I could not find them. My thoughts caught on that moment on the beach and the figures shadowed on the sand beside Mary and how she talked so earnestly to the air.

I was being ridiculous. Ghosts did not exist. The dead did not walk. If they did, would not Clara have come to me? Had I not begged for one last glimpse, a glimpse that would have shown me that, wherever she was, she was happy? No, bodies rotted down and the only claim to life they had was in the memories of those who lived on. My heart gave an uneasy twist. Perhaps it was my own steps upon the boards that had sent the engine rocking, and the sound that I had heard before, some strange effect of the wind. And as I took the stairs back down, I iterated it again. Ghosts did not – could not – exist.

*

I began to spend as much time as was possible out of doors. Away from Iskar, some of those feelings of anxiety were somehow temporarily relieved. Our hair would come loose in the wind and our skirts grow sandy.

Sometimes we just walked, in a silence that was never uncomfortable. The growing trust and affection needed no words, and slowly Clara and my father moved to a place that I could occasionally visit without such acute misery. And yet, behind all that, the sense of something astray sat across my chest and I was aware of a growing discomfort. '*All is not well*,' Ailsa had whispered weeks before, and I wondered if some answers might lie with her.

A few days later the opportunity presented itself and I found myself on the path to Ailsa's house.

I knocked and she was so soon at the door that I wondered if she had watched my approach. We went in through the kitchen, where the ceiling was hung with dried fish, game and herbs. A sharp, astringent smell was in the air, like nettle. She did not offer me tea or a seat but went straight to work stoking the stove, then pounding oats in a stone bowl. I stood awkwardly, playing with my coat.

'Sit, sit,' she commanded. 'You have no need to stand on ceremony here. You're not in Edinburgh now.'

I moved a pile of unspun sheep's wool from a settle, placing it carefully on the only remaining space on the table, and sat.

'So,' she did not look at me, 'you came to see me after all. How is Iskar?' She had strong features and low, falcon-like brows.

I framed my thoughts. Heat from the range warmed the room, and the scent of herbs grew stronger. There was something soothing in the rhythm of her strong hands.

'What did you mean by your words? What is not well?'

'I thought when I saw you that you looked like nobody's fool. Surely, you are learning their lives by now?'

'Of course.' I laid my hands on my lap.

'What do you know?'

'I would prefer that you told me what you were so keen to before. I cannot be disloyal to my new employer.'

At this, she laughed. 'There's no point having airs here, Miss Swansome. Why else did you come but to hear me?' She pushed the bowl away and regarded me with something like sympathy. 'I'm sure you're a good girl. You're not old either. What are you? Twenty? Twenty-two?'

'I am twenty-four,' I said.

'See, so young, and I can tell by your manner that you have come from privilege.'

'Even those from money can suffer some of the same griefs and misfortunes as those who don't,' I said.

'This is true and nicely put.' She sighed a little and looked at me more closely. 'You have been no stranger to sadness.'

I did not reply but the gentleness with which she spoke momentarily threatened to unlock my sorrow. There was a pause filled by the screech of gulls.

'I know about Hettie and William,' I said.

'What you may have been told, may not be the whole truth. I say this kindly. Hettie had a gift for second sight, did you know? She would have been wiser to keep it to herself.'

'For second sight?' I looked at her with surprise. 'I had not heard that. It sounds as if people believed her.'

A strange look passed over her features. 'She had not been here six months when she predicted a storm that would take a life. Two days later a gale stopped one of the boats trying to get back to safety and her prediction proved true. One of our boys drowned that day.'

I had not expected that and for a moment I was

shocked, but storms here were common enough and must surely risk the lives of those at sea. I did not believe that such foresight existed, and to use the vagaries of the weather to build a reputation for second sight was repugnant. 'That sounds like a prediction born of luck.'

She gave a small nod. 'Perhaps, and if that had been the only example then you could be right. There were others and then, not weeks before William's death, she foretold that too.'

'She predicted his death?'

Ailsa smiled thinly. 'Not so lucky.'

I tried to swallow my reaction, tried to cling hard to what logic dictated, but the sands of reason shifted beneath my feet. I could not find an easy explanation for such a prediction. 'Is that why the islanders considered her a witch?'

She looked at me with a glint of approval. Then she shrugged her square shoulders, covered the dough and put it on the range. Dragging out a chair, she sat and from a drawer pulled out a small pipe into which she put a twist of tobacco. Lighting a wick, she puffed furiously until it was lit. The smell was sharp and as unlike the sweet tobacco my father used as I could imagine. She blew out the smoke with satisfaction, daring me to pass comment.

'I think you know the answer to that question. You

told me you had learned what she was like. What she did.'

I nodded.

'She was clever with herbs, too. Women who talked badly behind her back, put their fear aside when their babies grew sick and sought her help. Some said she had hands that could heal.'

'These other abilities, if she really possessed them, could only have been assets, surely.'

She smiled. 'Hands that can heal may also be hands that can harm.'

I thought of the boy under her charge and what he became.

'But Miss Swansome, this is not what I feel I must tell you.' She spoke quietly now, as if she was afraid of being overheard, 'Do you know what role I take on the island?'

I shook my head.

'I assist the dead.' As she said this she looked to me for a reaction, but I kept my face impassive. 'Do you know what that means? I cleanse the bodies, wrap the dead flesh. If necessary, I bathe the wounds and correct the breakage of limbs. I place a coin on the tongue and sew closed the eyes and mouth.'

'It's a necessary job.'

'It's an act of love. When you die, Miss Swansome, would you not like to think that tender hands took care of you at the very end? That even in your coffin you were laid to rest as if you were preparing just for sleep?'

Her voice was like oil.

'What were you told about William's death?'

'He fell from Stack Mor, an accident.'

'But was it now?'

Her look was cunning.

'You do not scare me,' I said.

At that, she laughed again. 'I am pleased for that. I don't mean to scare you. I mean to warn you.'

'Of what?' In spite of myself, I shivered.

'William did not meet his death by accident.' Her words hit me with the force of a blow.

'I cared for his body. I saw the wounds he bore. And they were many. But only living bodies bleed fully. Did you know that?'

I was horrified.

'I took him and I bathed him. It is my belief that he was dead some time before he left the cliffs.'

My skin grew numb. 'If he did not die in the fall, how did he die?'

'There were many injuries, but in particular there was a strike to his head; it did not match the other wounds and it had clearly bled a lot. Some hours later his body was cast from Stack Mor.'

Dismay rose like an incoming tide. They had all

lied. No wonder his death was never spoken of. And there was Mary lost in the corridors of Iskar in a world of mute horror. Did she know too? Was this what glued her lips shut?

'Who did it? Has justice been served?'

'You think too well of people,' she said, 'or too ill. I'm no doctor, just a maid to the dead, and people do not like to believe such things can happen. Or, given his nature, maybe they did not care enough.'

And some of my feeling washed back with relief. Perhaps it was not what happened at all. This was one woman's word against a host of words. The island had not believed her and neither should I, if I so chose.

The afternoon was drawing in its colours and withering under the darkening sky. Ailsa took a breath of her pipe and let her gaze wander to the window. 'We're complicated souls, don't you think? Without the knowledge of who killed him, we are left only to imaginings.'

I had heard more than I ever wanted to hear and pulled my coat closer in anticipation of leaving. But she spoke again and the force of her look kept me fixed to the seat.

'There are other things that you should know, Miss Swansome. Listen.' She spoke so quietly that I had to strain my ears, 'When they brought him to me, I searched his clothing and I found something.'

I tried to resist the seeping sense of dread, but beneath the table I clutched my hands so tightly I could see the whites of my fingertips.

'Have you heard of a widows' whistle?'

The word slid through the air like something ailing, something I did not want to hear. I shook my head but my mind had already wandered to the sound I had heard from the attic.

'I found one in the pocket of his jacket.'

Although a part of me resisted, I had to know. 'A widows' whistle?'

'Do you know what that is?'

I shook my head.

'It's an instrument.' I could feel the pleasure she took in disturbing me. 'It's made of bone with dried skin for a reed. The first was made in grief a long time ago.'

I looked at her questioningly.

'A wife in mourning. A witch. She could not bear the pain of loss and called her husband back from the grave.' I could only stare appalled. 'For the best result, the skin and bone had to be human.'

I had heard enough now and burned to leave, but I could not move. I was transfixed by the curve of her voice. From the window, the falling darkness threw shadows between us.

'He returned – a ghost and so triumphant was she

in her success that she did not stop then; she called her lost children one by one.' She placed her hands on the table and trapped me with her gaze. 'A widows' whistle is blown to summon the dead. That is what it is for, Miss Swansome.'

14

I had been unable to broach what I had learned at Ailsa's with Miss Gillies as, when I had returned, she had been struck with stomach pain and taken to her bed.

In my room, I scoured the valley and observed the shivering treeline with distrust. Was it possible that William had been murdered? The island did not believe it, but I was more perturbed than I cared to admit. I told myself that the widows' whistle could not perform the task for which it was made; the ability to draw the dead from their graves with a pipe could not exist.

That is what I told myself. But as I turned to grab my shawl, I caught a glimpse of my reflection in the mirror and saw the thinness of my face, the lines of tension about my eyes, and I barely recognized myself.

It was with relief that I took the familiar path to the beach the next morning with Mary. The tide was out and the sea glittered in the scattered sun. Away from Iskar, some of my darker feelings were diffused, but from time to time, my thoughts snagged on what I now knew.

Mary walked further along the strand, leaning into a pool with her net. She looked back and waved and I found myself counting the weeks since my arrival and wondering what progress I had made. In spite of the warmth between us, she still would not speak and spent too many of her days clutching Bobbity, as if her life depended on it.

I became so lost in thought that I did not at once become aware that Mary was no longer with me. But I was not alarmed – the shallow caves where we often explored were just ahead. After a while, I put down my pail and followed the beach, fully expecting to discover her, but my calls and explorations brought nothing. A little impatient, I made my way back and then noticed how close I was to Gulls Cry and was struck with the conviction that this was where she had strayed.

In the morning sun, the house looked innocent enough, with white paintwork and wide windows, but its pretty face was a facade because I knew that behind the brick it wore another character; it was the place where Hettie had once lived alongside a boy with an appetite for torture. I began running to where the path led up through the dunes and on to the property. I was out of breath as I reached the door and tried the handle. It was unlocked, confirming my instincts.

I stepped into the lobby and the leaking scent of the stale rooms. 'Mary,' my tone was wheedling. Vacancy had set the house into fast decay – damp ran from ceilings and the scent of rodents was astringent on the air.

I paused, but a noise from above jarred me out of my reflections. I called Mary's name again and took the stairs to the first floor. Dust was distilled in the silence and I stopped to listen but did not hear her again. Pushing open a door, I discovered a bathroom – a hip bath, its enamel creped with age. Towels still hung on an ornate stand with the bloom of mould on their surface.

Her parents' bedroom was empty too but for a grand four-poster, hangings grey with dust, and the furniture that must once have been used daily by Evangeline. It was hard now to imagine the house filled with life or activity. Feet pattered in the hall and I left in search of her once again.

The sun must have been cast behind cloud because when I re-entered the landing it was steeped in shadow and an iciness had gathered. I walked a few paces to the next room. A pale light plundered the dimness but it was enough for me to see a row of toy soldiers that told me that this must once have belonged to William. I was curious, to view the space that had once belonged to him, to seek some sign of the things that I had been told. A quick search yielded nothing – only dirt and damp walls. As I turned to leave, my eye caught something beneath the dresser. I knelt and placed it on my palm. No larger than the pad of my thumb, cool and smooth to the touch and perfect in every way, it was the skull of a mouse. A combination of this discovery, the cold clamp of air and something in the hush made me uneasy and I left quickly, opening and shutting doors in my haste to find Mary.

It was as I stood in what must once have been her own room that through the window, I saw her – on the beach, a bucket swinging from her hand, her head turning this way and that in search of me.

I felt a moment of chilling shock. There was another shift somewhere in the house, and the boards groaned beneath a step that could no longer belong to her. I was not alone, and whoever it was was listening to me in the quiet.

It came to me then that Mary had never been in the house, that whoever I had heard had always been someone else. My neck shrank to iciness and it was as if everything unpalatable that had happened since my arrival at Iskar was rolled into a ball and pressed beneath my skin. I wanted to run but I could only stand, paralysed, and see her crouched over her stretch of sand. Although the world had momentarily stopped for me, birds still wheeled in the air and the tide crept unceasingly up the beach.

I could not breathe. I opened my mouth to call but the words were smothered in a sweat of fear.

There was another sound but I was already at the top of the stairs, steadying my hands on the rail. And then I raced down and outside to the open air and the cry of gulls.

Mary looked up when she heard me, the pail beside her. I reached her, breathless. My heart was a frenzy. Her eyes swivelled up to the house and to the room that had been hers. I twisted my gaze too. And for an instant, someone was there – the outline of a head and shoulders, the suggestion of a body, but whoever it was stepped quickly back into the shadow and out of sight. For a second my senses numbed and then my ear discerned a faint fluting. I shook my head but it remained – a whistling caught and carried in the wind. As much as I tried, I could not stop the way my thoughts were dragged to the instrument that Ailsa had described. A widows' whistle.

'Who is it, Mary?' I said. 'Who's in the house?' My voice fluttered like a moth caught in a jar, 'Mary, answer me.'

Her eyes met mine with intensity, then she crouched, her hem catching a puddle of sea, and with her finger, she began to write in the sand. As quickly as she did, each indent filled with water and was swallowed, but I kept the letters in my head, and even though the name had gone it rang in my skull like an alarm.

'William', she had written.

William.

15

I do not fully recall now the return to Iskar or the weight of discomfort that sat inside me. I know that more than once I turned to look back at Gulls Cry, to see if someone stood at a window or left by the door, but the house lay still and heavy like somewhere long unoccupied.

I know that I pondered again the whistle of which Ailsa had spoken, thought of what it was made for, remembered the sound of childish steps in Gulls Cry and William's engine on the floor of the attic. I had to stop in my stride to catch my breath. Looking up to Iskar, part of me withdrew like a crab into its shell. If I were to believe that such things were possible, all would be lost. To what would I cling? And so I pressed that dreadful idea back out of sight and reminded myself of what we knew for certain about the universe, that night followed day, that age followed birth and after death . . . after death, there was perhaps a place with God, or more likely the silence of an abyss.

In spite of my resolution to be calm, dismay made a fist in my chest, and as the morning progressed, a headache grew. I did not have an appetite for lunch and instead went to lie down. Mrs Lenister brought me up a hot drink laced with herbs. I longed for someone to talk to.

'See if you can sleep,' she said and pulled the covers up around me, tucking them in almost as a mother would. The gesture touched me nearly to tears. My thoughts returned again and again to the morning. Had Mary believed that she had seen William? Had it been an island child and perhaps Mary missed her own brother so much that she conceived his face from another's? I recalled those early days when I was convinced I saw Clara everywhere – on a crowded street, or the back of her head a few pews in front at church, and it was only when I looked closer that I realized it had not been her at all.

I remembered the sound of the feet at Gulls Cry, their weight and tempo – whoever it was had not been adult; the steps I heard were those of a child.

The drink was hot and bitter and I slept for several hours, waking to sparrows on the sill. My dreams had been embedded with fear and the image of William's body upon the rocks in a ring of pebbles.

Getting up, I made my way out of the room. The corridor was hushed. Shadows fell from the attic and pooled on the threadbare rugs. I listened to the dull creak of the eaves. Beside me was the door to

William's wing and for an instant I thought of his palm upon the handle, a monstrous token of the day's work in his pocket and a whistle on his lips.

Instead of making my way immediately to the drawing room, some compulsion had me take to the attic where threads of icy air spun in the atmosphere. I was aware of my human weight on the treads, their reassuring firmness, the way my heart beat inside my living chest. The mind, my father had always taught me, was the most unreliable of man's organs. Hearts could fail, kidneys could grow septic, but the brain was capable of further reaches; we trust what we see and hear, but how do we truly justify the conclusions we reach?

Hettie's room was cool on my skin. I pulled the curtains open, flinging a pale light across the fittings and furnishings. It was silent but for wind and sea. No engine rattling the boards, no figure standing in the corner. It was empty. I had a sudden image of Hettie far away under the American sun, swinging her hips with a child on each hand, a secret smile about her face. The way the children would cling to her, unknowing of how close their enemies were. Wherever she was, she was ignorant of the fact that not weeks after she stepped on to the boat William lay dead at the foot of Stack Mor, rendering his sister dumb.

A gust of wind found an aperture in the window

and grabbed the curtain, throwing it against the wall and causing a slight banging. I recalled a trick of my aunt's and went to explore the hem where some of the stitching was loose. As I felt the fabric, I came across whatever it was that had been hidden there. I widened the gap in the seam. Inside was a brooch in the shape of a dolphin. The jewel of its eye shone in spite of dust. Why would Hettie have been so keen to keep this from the reach of scrutiny? It was not a cheap object, yet she did not value it enough to take to America – or had she forgotten it in the haste of her decision to leave? With it was a letter.

Dearest Hettie,

Promise me you will always be mine.

I recalled what Mrs Argylle had said about Hettie's various alliances and wondered who this had come from. The island men did not have wealth, as a rule. Carefully, I put the things back.

As I left, there was a beat of sadness upon the air. I tried to dismiss it but it came stronger, pressing at my heart until I could not move. I turned, and for an instant I thought I saw her in the dressing table mirror, her face twisted with grief, but it was so soon gone that I was not sure it had been there at all. Even so, I ran from the room and to my own where I waited for my heart to stop its frenzied beating.

In the drawing room, I sat in the faded grandeur and waited for lessons to finish. I heard them before they entered, Miss Gillies' clipped shoes upon the tiles and the shush of Mary's slippers. And then Miss Gillies' voice like the lash of a whip. 'Wicked, wicked girl,' she hissed. But I was unsteadied by the morning's events and had no energy left to defend my charge. I did not want to be found, did not want Miss Gillies to know that I had heard her chide her niece, and as their footsteps faded, I fled upstairs until tea time.

Mary was alone when I returned to the drawing room. 'I will have to speak to your aunt about Gulls Cry,' I said gently. 'An intruder must have opened the door.' Her brow rose in a spasm of anxiety.

'I will not tell her that you wrote William's name,' I said, 'but if you saw who was really there, you might write it for me now.' But even as I uttered this my thoughts drifted back to the sound of young steps and the noise of an engine being pushed across the floor of an empty room. And my words felt like a betrayal.

I found Miss Gillies in the morning room. Candlelight glittered on the walls and bounced off the jewel she wore in her hair.

'When Mary and I were at the beach this morning,

we discovered the door to Gulls Cry open and I believe someone may have been inside.'

Her mouth creased with displeasure and her head seemed to drop a little on her neck. 'Thank you, Mrs Lenister informed me. I suppose it's easy enough for anyone to break in – it wouldn't be the first time. I shouldn't be so surprised that some child or other has gone to explore, maybe taken a trinket or two, but you're right to be concerned, the disrespect upsets me.' She sighed sharply, causing the candle's flame on her desk to dip.

'What will happen to the house now that Mary is with you?' I asked.

'The house was given to Evangeline when she married John. My father had it built, just for them. When Mary is old enough and married, she will inherit it.' Her tone was pinched.

'And you were always at Iskar?'

'Always. Once, I had thought that I too may leave but that was not to be.' She looked around her. 'But Iskar is a fine house. I love it as well as I might a husband,' she said dryly.

I could not keep quiet. 'Miss Gillies, I was told a rumour yesterday concerning William's death.'

She looked at me wearily. 'I heard that you had been to visit Ailsa.'

I flushed.

'Ailsa is a scaremonger. I imagine she's been waiting with bated breath to whisper a lot of nonsense into your ear from the first moment of your arrival. The rest of us have learned to pay her little heed. You would be wise to do the same.'

'But she said . . .'

'I know what she says,' her face was strained. 'She says that William was murdered. I can tell you now that he was not.'

'He was not?'

'Absolutely not. The cliffs are hazardous, even in clement conditions. It was nothing more sinister than an accident.'

I looked out of the window and to the gardens where bracken had taken hostage of the beds and overgrown rose stems drooped with frost. I thought about Ailsa's manner and the enjoyment she took in unnerving me. Miss Gillies' confidence was the reassurance I needed.

Her eyes went to a portrait on the wall. 'I have thought a lot about my sister recently. She was such a character. See,' and she pointed. 'Come, look at her likeness.'

We moved closer to the picture and she held her candle up. I studied Evangeline's broad forehead, lips that were fuller than her sister's, and I could see the ghost of what Miss Gillies might have been without the scarring. What must she have felt as she looked at her sister's unmarred face? It must have been a constant reminder of her disfigurement.

'Did you have other siblings?'

She shook her head.

'Just one younger sister?'

She looked at me then. 'She was not the younger. I am the younger.'

I was hit by a faint shock: why had Miss Gillies inherited Iskar and not Evangeline?

My thoughts stumbled but I reined them back. 'She's very lovely,' I said. 'It's a shame that you were not closer.' Almost imperceptibly, her fingers crept to her cheek; her lips were white, almost bloodless. A spill of anger was in the air. I opened my mouth to say something but she was looking with such intensity into the lines and shadows of that beautiful, undamaged face that I held my tongue.

Later, I took Mary to her bed. She was too tired even for prayer and I helped her into the sheets and pulled the covers to her chin. For a moment she lay there, her eyes on mine, and then she reached out and touched a curl of my hair, twisting it in her finger. The warm pad brushed my skin, as gentle as a feather, then she moved her hand down to turn the bangle on her wrist with a look of pleasure. I kissed her cheek, inhaling for a moment the warm scent of her breathing skin.

The darkness was brushed with the faint bloom of a moon and the persistent rush of the sea. I sat on the chair and read to her until finally she was asleep.

I could not push the morning from my thoughts. I pondered on Miss Gillies and the staff – the silence that followed the mention of William – but I felt him everywhere – everywhere, and that name written in sand, it seemed written on my fears.

When Greer came with her hummed lullaby, I was weary of her. Go away, I thought, I am tired. But she came anyway and my skin flared awake.

When her voice dropped entirely, my ears rushed with silence as if a shell were held to them. She stayed until my throat ached to call and then, almost like the break of a wave, she sighed and her steps faded as she continued on her way, the lullaby rising and eventually falling till it receded beyond reach.

16

Mary had woken me in the night with another nightmare. It had been a long time since Miss Gillies had taken an interest in Mary's nocturnal dramas, which at least was some relief, and she had them frequently enough. This time, Mary had taken a while to settle and afterwards I had lain awake, filled with unease.

I could hardly wait the next day for lessons to begin; time seemed to drag, but eventually Mary and Miss Gillies left for the schoolroom. Outside, the sky was white and salt spray soon wet my lips. Smoke rose comfortingly from the Argylles' chimney.

'Miss Swansome, what a lovely surprise. Sit whilst I arrange for tea. I thought that you might visit.'

'I hope that I have not come at an awkward time, Mrs Argylle.' I could not hide the rush of emotion. My desire to speak frankly battled with the responsibility I felt for not taking gossip of Iskar outside its walls, but it seemed that William's nature was well enough known. The maid came with the tray and we waited until she had left.

'Please,' she said with a smile, 'we are friends now. Call me Bridget. I can see you are troubled.'

And all that I had recently learned about William fell from my lips.

'You have heard it all, then, and I know if we discuss it, that I can trust you to be discreet.'

'Of course.'

'I was Evangeline's friend. I knew William well and it was always my opinion he was unnatural, even before Hettie taught him what she did. I think he was one of those children one rarely sees – there was something different, something missing.' She paused. 'Love. He was without love of other living things. It had turned itself inside out. And instead of love there was contempt and anger. All Hettie needed to do was nudge it further in the wrong direction.'

I had not wanted to hear confirmation. I had hoped for something else, some knowledge that would have lessened or cast doubt on the charges laid against him.

'They were both strange children,' and I caught an inflection there that I did not like, 'but I think William was born clever.'

'Born clever?'

'Some souls are made to be dark.' She studied me with something like pity. 'The world gives birth to both the viper and the lamb, and there are churches for each.' Her voice was almost a whisper now.

'And Mary? You said they were both odd. You cannot think that Mary also took part in this?'

The lines about her eyes tightened.

'No, of that, she was never witnessed, but do you think Mary natural in other ways?'

I thought of the name she had written in the sand, the way at times I caught her gaze at some empty spot as if in recognition, but I did not want Bridget to think badly of her. 'Yes,' I said with vigour, 'I do. She has a loving heart, I see it all the time. It's just grief that has rendered her to the state she finds herself now.' I was shocked to discover that my eyes stung.

'I see that I have not eased your discomfort but made it worse. Hettie is gone and William lies dead. Try not to let them concern you.'

'As to his death, I understand there are those that say he was murdered?'

She raised her eyes. 'You can't think that's true?'

'No, I understand it's not commonly thought, but it occurred to me that his badness would be a motive.'

'Skelthsea may have those who believe in the old ways, but there are none here who would consider death a suitable punishment for sin, and certainly not for a child.'

I was further reassured. 'There are other things too,' I said cautiously. 'Ailsa told me of a whistle that she found in William's pocket. She said it was to summon the dead. And Iskar, in truth, feels at times as if it is haunted.'

Bridget's face creased to laughter which, seeing my distress, she quickly controlled. 'You must not let Ailsa scare you. She loves to spread mystery. You surely don't think such things?' And behind her kindness I detected a faint note of incredulity that I might have allowed myself to consider that the whistle could possibly have the power to do what it was made for.

'Of course not,' I said. 'Ailsa unsettled me and it's hard to think of a child in his grave, whatever his crimes.'

She laid her warm palm over my hand. 'You are a kind soul.' And I was shocked to see the glaze in her own eyes.

'You asked me once if I had had children. I carried three, you know.' The strength of her emotion pulled at me. 'No, not to birth. I lost them, lost them early.'

'I am sorry,' I said.

'We wanted them very much – Robert as much as I – oh yes, we wanted them.'

I held her hand tightly.

'See,' she said, 'I have not spoken of this other than to my husband, but a burden shared is not always a burden eased. It doubles our agony, but you – something in you makes it possible to say what I have said.' She leaned over and kissed my cheek. 'You comfort my heart. And do not let Iskar unnerve you.

'I think if I had to live there that I too might feel the same. It's always been a place that summons a fertile imagination, all those dim corridors with creaking wood and wind in the chimneys. Too easy to imagine a sound or catch a shadow and turn it into something else. Iskar can feel very empty with so many unoccupied spaces.'

But all I could think of then was all the unoccupied spaces that did not feel so.

Not long after, I thanked her and left.

Outside the sun was beginning to sink on the horizon. The grasses bent against the wind and I shivered. My eyes found Iskar, where somewhere Mary's young head would be bent over a book or a piece of work. A child who had lost everyone close to her. I pondered how those losses must feel. How little comfort she had received from the woman in whose care she had been left.

I tried to imagine her grown to full stature, tried to imagine her face with a womanly cast and how she might sound, how her walk would adopt the elegance of a more mature figure, but I realized that I could not. I paused, the cold air clinging to my skin, waves slamming the rock-line, and I was jarred by some deep horror. All I could see was Mary, sunk in the oblivion of water, her white limbs waving on the pull of the tide as she sank deeper and deeper into death.

As I neared Iskar, I tried to quell the anxiety of the image I had seen of Mary – surely something born of the recent talk of William's death – and yet, as much as I dismissed it as fanciful, I could not fully shake it. I remembered that first afternoon, the way I had sat in the hall, balancing the weights of expectation and loss – how I had anticipated Mary. Never for a moment would it have occurred to me that I would be entering a life that would take so many dark and strange directions and a house with such a presence. Although I had now been told by three people about William and his wickedness, a little part of me still resisted. I had to know it for myself.

Later, as we sat at tea, Miss Gillies questioned me about my visit to Bridget, but all I could think of was the room locked behind the door to William's wing.

Mary sat silent with a book, only looking up when I spoke, and I was struck anew by how distant their relationship was, how little Miss Gillies addressed her niece or spoke a kind word.

Finally, I took my handkerchief and coughed into it then laid my head against the chair.

'Are you unwell, Elspeth?' she said.

'Only a little chilled, but I think it best to stay when the rest of you attend chapel.'

I lingered at the drawing room window, watching

their retreat. When they were on the village path, I went quickly to the kitchen where I found the keys for William's wing.

My heart pounded as I made my way along the hushed corridor. At the foot of the staircase leading to the garret, I paused and glanced up to the old nurseries, but I did not linger and turned the key.

Once in, I faced a passageway similar to the rest of the house, although the scent of damp and decay was worse. Most rooms were empty. It was at the end that I discovered where William must have slept.

The room was sparse. I rifled among the clothes and objects, curious to find something that would connect me to his reputation, but the search yielded nothing. I realized that I was grateful there was no evidence of his badness and wondered if his reputation had been exaggerated.

Time was pressing; the cold would surely keep the service short. I sighed and because my feet were growing numb took one more circle of the room. It was as I made my way past the window that I was aware of the way the floorboard creaked differently beneath my foot. Unlike the others, this was not nailed. I rolled up the rug, saw the faulty joints and pressed my thumbs to a corner. It lifted.

Something inside me thrilled but was repelled when I realized that amongst the dust and cobwebs lay William's secret collection. I held my breath, half-dreading, half-fascinated by what it might contain. Laying the objects out removed any last trace of doubt. Why else would a boy have need for such things: the skull of a bird, bone pale and beak sharp, a knife, curved and thin with a silver handle? I did not want to pick it up; in the weak light it was clear that the blade was rusted with blood. These artefacts told me William's story more vividly than words had done.

Amongst the belongings was a leather pouch. I tipped the contents on to the boards. Repugnance rattled in my chest. I knew immediately what it was – what it might have been had he finished it – a widows' whistle. As Ailsa had described, the bone was roughly hewn and beside it a membrane of dried skin from some wretched creature. He had planned to make another. My knowledge of biology was too thin to confirm or deny whether the bone was human, but I was shaken more deeply than I could have anticipated.

My knees were unsteady as I walked blindly to the window and rested my palms on the sill. Against my will, I was revisited by the sound of feet at Gulls Cry. I gripped harder. No, I told myself, just because he believed or Hettie believed, it did not follow that it was true; the whistle could not hold the power it claimed. And if I thought I heard his toy or for an instant glimpsed a reflection, how easily could I have misread it? And as my heart began to beat more calmly, I returned to his store.

There was little left now: candles and finally a collection of stones, stones that were familiar – marked with a figure and bound with hair. It was only as I put them back that I noticed the small box tucked in a corner.

Inside was a doll such as the one I had discovered in Hettie's room. And as that one had been, this too was without a face. Next to it was a handkerchief stitched with an initial – H. Hettie.

And I imagined them together, he and Hettie, trapping the creatures, crafting bone to call the dead, and I was visited by rage for the woman who had manipulated a child so. I fixed the boards back into place and stood shivering in the chilled air. They had not lied or exaggerated. William had gone bad to the core.

Outside his room, I walked to the end of the hall and came to another staircase leading to more attics. The steps were worn and woodworm had dropped dust to the treads, but there were cleaner patches where someone had placed their feet. When Greer opened this wing, was this where she went? Did she know of William's hiding place? I was keen to climb but the hour was growing close and I knew that I must return the key. I did not trust Greer's sly looks.

I felt in everything she did that desire to make mischief. I turned to leave, pausing once more in William's room, and saw something that I must have missed on my first search, yet I could have sworn it had not been there before. Laid so carefully on his bed was the red engine with the rusted wheels.

I left the wing at a run. Outside, even with the door re-locked, my heart thumped and it was as if the tainted air had settled on my skin. I put my palm out to touch the wood and leaned close. And thought I heard it.

A distant sound, thin and wavering.

17

As Mary and I walked on the sands the next day, I gazed across the ocean to the home I once knew and was aware of a deeper dismay. As if sensing some withdrawal, Mary kept close and her eye was on me often as if searching for some proof that I cared less.

Her nearness was my only comfort. And if I had any doubts, I told myself that there was no evidence that she too had turned bad. I would have seen or heard of it by now if that were the case, and other than the pebble I had found in her pocket, there had been nothing else to hint at aberrant behaviour. Even so I was aware that my hand might not wield as much influence over my charge as I hoped.

How I wished, then, that I could catch a cab to some cheerful market town – full of the smells of cakes and cooking meat, stalls of flowers and costume jewellery to engage my eye, a coffee house that sold fancy cakes and chocolate and a hot, sweet drink to follow. I recalled what it was like to hear my heels clipping on cobbles, to smell pipe smoke, the gas of the streetlamps and factory fumes, the belch of alcohol and the swell of sound that emerged from ale houses.

Here, there was nowhere to go and only a monthly boat in winter – if we were lucky.

Mary went on ahead, her thin legs throwing long shadows across the bay, and I brushed away my thoughts and caught her up, tapping her shoulder and running off so that she could catch me in a game of chase.

Before lunch, I went to my room and found the pebble and put it in my pocket. As much as I wanted to confide in Bridget, I was aware that I did not want her to suspect that Mary was the one who had been in possession of such an object. It was clear she already believed Mary strange and I was keen not to add fuel to those doubts. Even so, I had a burning desire to learn more, to know their meaning; if William and Hettie made spells, what was the purpose they sought? Perhaps in understanding that, I might understand more of why Mary had kept one. It was to Ailsa's I planned to go.

I was impatient for lessons to begin, and as soon as I heard the door close I made my way back outside. When I reached Ailsa's, I knocked hard. When she greeted me, there was a smile in her eyes that told me she had expected me back.

This time, she offered me a seat.

'And so?' she began.

I drew out the pebble, placing it in front of her.

She pulled it to her curiously. 'Where did you get this?'

'I found it in Hettie's room,' I lied. 'What does it mean?'

She opened the drawer, took out and filled her pipe. 'I think you know the answer to that. It's a spell.'

'What spell?' I asked.

'It's binding magic.'

I looked baffled.

'Binding magic is made to attach one thing to another – but here,' she touched the pebble, 'this figure tells me that it was made to bind a person to something or someone else. When it is a person, for such a spell to work the hair must belong to that person too.'

'But what does someone achieve by making it?'

Ailsa leaned back in her chair, blowing smoke into the air. 'The idea is to tie the person to something for eternity. Commonly, it was made to bind one to a lover, but it could also be made to keep one tethered to a place.'

My chest relaxed. 'This, then, is to bind someone to another? To make them love them and not leave them?' And I ached with remorse that I had considered the pebble might have a more sinister meaning.

Who more than Mary might long for that to be possible?

She turned the stone over in her palm.

'Whose hair is this?' she said, holding it to the light.

'I don't know. It's of such an indeterminate shade, and probably dirty.'

Putting it down she reached across and took a strand of mine in her fingers. I could smell the dry smoke of her breath. 'You say you found it in Hettie's room?' Her eyes were sly. 'It could almost be the colour of yours,' she said.

My eyes flew to the stone and yes, I saw now from the streaks of darker hue that the hair could indeed have been mine.

I took the pebble back. 'Thank you,' I said, preparing to stand.

But she clutched my wrist. Her hand was warm and rough like a man's, her eyes troubled.

'There are other binding spells.'

'Others?'

'These spells can bind you to love but they can also be cast to bind the person to illness or death.' She watched my face and I felt that she knew that I had lied about where it had come from.

I had to look away to conceal how much her utterance appalled me.

'I don't believe in witchcraft,' I said with more

confidence than I felt at that moment, 'only that some might believe themselves capable of carrying it out.'

'Then why does this concern you?'

I swallowed, feeling trapped; she had found the chink in my reasoning. 'I thank you for telling me,' I said and tried to pull my hand away, but she held it fast.

'This was not where you said, was it? Do you know who made it?'

I shook my head and wondered if she could feel the pulse that beat so hard beneath my skin. 'It was in a drawer in Hettie's room,' I insisted and she let me go, her eyes bright with disbelief.

At the door, she studied me again. 'You should be careful,' she whispered. Her eyes bored into mine and I felt her plucking my secrets like fruit from a tree. 'But you know that already, don't you?' I only nodded and stepped into the cold.

I was grateful for the air that pummelled my body. The muscle of the wind was in the bent trunks and in the twist of birds. Below, the waves rose and grabbed at the shoreline and battled with the rocks. There was fury everywhere. Clutching my hat, I raced down the valley and back to Iskar.

And then we were walking up the path to the chapel and its familiar smells where we shuffled into the pews, the smell of wax smudging the air. I glanced at Mary and at the space that would remain forever empty beside her. The rustle of the Bible brought me back to the present as Robert Argylle began to read, but I struggled to concentrate, losing my place in the sermon.

I did not want to talk to the villagers and so I hurried down the path. At the strand I met Reid Paterson. I started and he moved to stand before me. His hair was slick beneath his hat.

'Good evening, Miss Swansome.' His eyes roved my face with rude intimacy.

I nodded and sidestepped him to pass, but anticipating the manoeuvre he did the same, leaving me trapped.

'You've become a stranger. Has the mistress warned you away from me?'

'I have had no need to make any purchases.'

He laughed. 'Oh, I think we both know that that has nothing to do with it.'

I pulled my coat closer. 'There can be no other reason to visit your premises except that I am in need of something.'

'There are other things that young ladies with no suitors can gain from me,' he smiled, 'and they would be free.'

I felt a rush of indignation. 'I have no need of

anything that you can offer and cannot imagine a time when I will.' I went to move away but he stopped me.

'It must get a little lonely, just a house of women. All those empty rooms.' He reached out a hand and plucked my sleeve.

'I am quite happy,' I said, pulling away angrily. 'Excuse me.'

His teeth gleamed and his fingers closed around my wrist, forcing me to look at him. 'There is no point taking on airs, miss, not here. I know your type.' His eyes drifted to Iskar. 'Your predecessor was not so unwilling; indeed, she was not the only one.'

Hettie? Was Paterson one of her alliances? I thought of the brooch and its value. Paterson was one of the few island men who had the kind of wealth to purchase it. I pictured them together, saw the commonality of handsomeness, and knowing Hettie's nature it was too easy to imagine that they were lovers. I tried to pull away and he laughed, 'I'll be waiting. You know where I am.'

A noise behind caused him to remove his grasp and I breathed out with relief. Greer stood there in her cheap hat. Her eyes looked from me to Paterson. At this Paterson gave her a smile, a gentler one. Then he backed away, gave us both a bow, and we walked on.

'I was glad that you came when you did, Greer,' I said.

She did not reply.

'He is very forward. I feel sorry for his wife and children. And right after church. I had heard that Hettie made some affiliations. Was he one of them?'

She glared at me and bit her lip, but there was a flare of knowledge in her look that made me suspect I was right in my assumption. We turned to ascend the path to Iskar. Greer went round to the servants' door and we met again in the hall.

'You watch out for him,' she said. 'There's no compliment to your looks there, if you get my meaning. He has tried his hand with every woman on the island, ugly and pretty alike. Perhaps he thinks he'll be luckier with someone like you.' Giving me a cruel smile, she took my coat and disappeared behind the baize door.

Greer's put-down grated and I hurried upstairs to change. As I came down afterwards, I encountered Mrs Lenister.

'Blowy afternoon,' she said. 'Are you quite well?'

I sighed. 'Quite well, thank you, but I do not seem able to please Greer today, somehow.'

'Ah, Miss Swansome. Don't mind her ways.'

But she could see from my expression that I did. I turned to leave and then recalled the shoes that were now moved every night.

'Every morning I wake to find that Greer has moved my shoes and placed them oddly.'

'Moved your shoes?' Her eyes went a little wide. 'She has to move them, if you don't mind me saying, in order to clean.'

'You misunderstand me. It happens at night,' I said. 'She must come while I'm asleep and place my shoes heel to the wall. Every night. I do not understand why she does this.'

Mrs Lenister opened her mouth but no words came.

'Is it a tradition here?'

She shook her head but her skin paled. 'No. It's not something that we do here. Not at all.'

I felt her withdraw, a shift in her energy. 'Excuse me,' she said and swept past me to the hall beyond.

Mary and I were served dinner in the playroom, but my head throbbed and I left Mary to go and lie down. It was not long till I fell asleep, waking later to the weight of darkness. The fire lay low in the grate. It was bitterly cold. Chilled, I got up, tidied my hair and dress and went to the playroom.

Inside, there was whispering. I opened the door but Mary did not hear me. She was kneeling intently over some pursuit, her quiet words punctuating the silence.

I stepped softly towards her, curious to catch what she said. But the language was the unintelligible one from her nightmares and the part of me that had been sparked by interest, shrunk. When she finally became aware of me she snatched at the ground but was not quick enough to hide what she was doing. There were five pebbles laid in a circle; each had a figure drawn upon it, each with hair wrapped around it.

I was cut with dismay. Although I had suspected that she might have some understanding of the stones, it had not occurred to me that her interest had extended so much in their direction. Did she play with them in full knowledge of their meaning?

As if sensing my consternation, her face lined with anxiety. I knelt and touched one, picked it up and studied it. The hair was dark or dirty. It could have been any colour, even red beneath the grime. I walked to the window to give myself a moment of calm. Behind me there was silence but when I turned, tears ran from her lashes and on to her face. She wiped her nose with the back of her hand and watched me, fear dilating her pupils. Light from the glass caught the slight pink of her pale hair. Again, I reminded myself this was a child in pain. If she thought that the pebbles could bind one to love, who more than she, who had lost so many, would be drawn to that idea? Even so, it could not continue.

'There will be no more of this,' I could hear the height of emotion in my words.

She nodded.

'I want to understand,' I said.

She met my gaze but with that strange, removed expression.

'These stones, Mary, the ones with figures on them. What do they mean to you?'

I thought she whitened but she refused to meet my eye again. Rising to my feet, I took some paper and a pencil from her art box.

'You must tell me,' I insisted, in a tone that I had not used before. I wanted to hear her say that she did not know or that she had made them with the purpose of love.

Although she tried to hide it, there was distress again in her look.

'Please tell me, Mary. I want to understand. I know your brother was not a good boy. I know it and you know it too, don't you?'

She flushed and pulled the paper to her and began an angry scrawl. When she had finished, her breath escaped her in a gasp and I took the page from her.

'Death', she had written. 'Death. Death. Death.'

With each word the force of her pencil scratched deeper into the paper – then she sprang to her feet and left the room, slamming the door behind her.

18

After trying to comfort Mary the previous night I had returned to my room, but the experience had left me numb with horror. I woke to the image of her beneath the water and the words she had scrawled on the piece of paper. In the mirror my skin was white, my eyes haunted. The day could not have been a good one, however much circumstances might have made it so. Today should have been Clara's birthday.

The picture of the two children and Hettie was propped on my dressing table, but I did not need to see William and Hettie's images to recall the detail of their faces; they were branded on my unease.

As I splashed water to my cheeks, Iskar itself seemed to watch me from its shadows, seemed imbued with death and all that was wicked. It was there in the air, in the ailing whistle of wind in the pipes and the unnatural silences that occupied the gaps of sound.

Mary was subdued when I woke her, the sky outside a smoky grey. I had collected the pebbles the night before and placed them in a bag. I needed to be rid of them, but as I brushed through her hair it seemed a small gesture.

It was spitting rain but Mary did not protest when we left the house, and soon we were on the sand, the waves tossed with spume and a mist rolling in fast from the horizon. Gulls mewled on the air and I sensed the coming of a storm.

Occasionally, Mary paused to pick something from the ground – a shell or stone – and Iskar fell further behind us. We passed the caves and came upon the wide ledge beneath the cliffs. The smell of seaweed and brine was strong. Waves mounted the plateau, but holding to the rocks, we continued on until we were in the force of the wind.

I drew the bag of pebbles from my coat and took one out. Mary watched as I lifted it up and, as an afterthought, unbound the hair from the figure and let the wind rip it from my fingers and then I threw the stone to the ocean. I did it again and again until I came to the very last. My palm wrapped about the cold surface but I stopped and paused, something in me inexplicably resistant, and instead of disposing of it, I returned it to the bag.

We made our way back towards Iskar's cold reach. A row of gulls preened on the roofline, but my eyes were dragged upwards to the attic where shadowed behind the glass someone stood and watched us.

I stopped. Mary was ahead, her braids and coat tugged in the gusts. I tried to read from the indistinct shape who it was but darkness blurred the edges. Was it Greer? I could not tell. I did not move but stared back until my eyes burned, and it was only as we reached the path that the pane lay empty.

I felt it then, a sense of loss, of defeat. In spite of my efforts, Skelthsea was beginning to slip away from me. I no longer saw the shining curve of the bay or the sensuous sweep of the valley; once again, my eyes sought out the land beyond the ocean to the place that had once been my home.

The hours passed uneasily to an evening crushed in silences and my feelings began to thaw, leaving torment. From time to time, my fingers went to my locket and as I talked and ate I could only think how, but a little over eighteen months ago, Papa and Clara had both been alive and life had seemed so easy. The ache grew sharper, pressing upon guilt and anger. When I took Mary to bed, she held on to my hand as if to make me stay, but I was too weary of being awake and wanted only sleep and to close the door on such a day.

In the solitude of my room, I allowed my memory to drift back to this same day, but a year earlier, when I had stayed up late to decorate the little parlour with fern and dried physalis stems. The presents had lain wrapped on the table by Clara's chair. Once again, I was by the stove with the hot blast of air to my cheek as I had sat each evening to finish the dress in time for her birthday.

I could not stop my remembrances even when they returned again to the charred air, and the roaring flames that stole the sound of birdsong; how I had launched myself through the blazing door and into the melting heat, pressing my dress to my face, each call choked back in the maw of fumes. Lying flat to the hot stone, I had pulled myself into the furnace, terror wrapped like a drum skin over my heart, but I had known already that it was too late. I had known in my soul that Clara was dead.

Alone now, in the coolness of my room, I undressed and lit a candle, took my locket and gazed for a long time at the tiny image. With pain, I saw that it had faded a little and my chest burned with emotion. It wasn't until I got beneath the covers that I found it – there, on the sheets, a piece of paper and a small sketch of two smiling girls. It was clearly drawn by Mary and underneath she had written, 'Elspeth and Clara'.

Mary had noticed my pain. She had seen it even though I had not spoken, and my heart ached for her kindness.

19

The following day, even the salt-spit and winter sun that cracked the ocean did not banish the lowering of my mood.

In spite of the recent unpleasant interaction with Paterson, I went to the shop and bought toffee which Mary and I then ate in the lee of the rocks. Occasionally, we turned and met each other with a smile at the gluing of our teeth. But the smile was only skin deep. I had lost not just confidence but hope.

Back at the house, I sent Mary upstairs and went in search of Mrs Lenister.

'Would it be possible to have tea sent up to the playroom?'

She put down her knife. 'Certainly, Miss Swansome. You will need something hot in you both after being out. I can see the wind in your cheeks. They couldn't be much redder.'

I took my time returning to Mary and when I arrived at her door, I paused. I remembered that first morning when I had imagined William racing the corridors of Iskar in a game with his sister. No such imaginings remained – instead, I saw him crouched over the boards in his room to add some gruesome trophy to his store.

From inside the playroom, there was whispering. I swallowed my unease and put my ear to the wood. I did not want to go in lest I find her again in some pursuit that I was not ready to digest. So I stood for a few moments before turning the handle. She was facing away from me. Light flooded in from the window. Bobbity swung loose from one hand, but as I grew nearer, her words seemed intense and directed and I saw her in the pane where her face was twisted with anxiety. Lifting a hand, she bit at the edge of her nail and fell silent. Then she began once again, quieter now, and I realized from the rhythm that it was as if she were talking not to herself but to someone else, and I was reminded of the day on the beach. I inched forward. My new position gave me a better view but she seemed to be in shadow even though light fell cleanly through the glass.

And then, for an instant, I saw it. I blinked, looked again and instead of looking at Mary I studied the space beside her. There, just to the left of her, was a fainter outline, a little taller in height than Mary. I dropped the gloves that I was holding and gasped. Mary lifted her gaze and whipped round. I opened my mouth to say something but nothing left my lips.

She took one look at my expression, clutched Bobbity and ran past me and to her bedroom where she slammed the door, leaving me standing in the cold.

My eyes strayed to the window once more, my heart beating heavily. The glass was clear now, there was nothing there, but, by a shift in air, I felt that something other had moved to stand beside me in the quietness. In my reflection my skin paled. Then, with the faintest change in the atmosphere, it was gone.

I went into the corridor and shut the door behind me, leaning against the wall for support. Footsteps began from further away, moving closer, and soon Mrs Lenister's cap rose above floor level as she reached the top stairs. In one hand she held a tray. Straightening my dress, I stood tall and tried to smile.

'Are you all right, Miss Swansome?' Her brows drew together but her face softened. 'You look awful white.'

'I'm fine, thank you, Mrs Lenister.'

'A hot drink will set you both up on this chill day. Is Mary in the playroom?'

'Bedroom.'

'Go and sit by the fire. For a minute it looked as though you'd seen a ghost.'

Something must have faltered in my expression and her eyes narrowed a fraction. She followed me into Mary's room where Mary lay upon the bed. Mrs Lenister did not comment but placed the tray on the table and prepared to leave. My feelings stuck in my throat and on impulse I went after her, catching her as she was about to descend the stairs.

'Mrs Lenister.'

She turned questioningly.

'Ghosts,' I said. 'Have you ever . . . has . . .' I did not know how to finish.

A flush rose on her skin and she held herself a little higher. 'This is not something of which you should talk, Miss Swansome.'

'So, nobody has ever experienced or spoken of anything here?' I pressed.

Her eyes shifted. 'If you'll forgive me giving you a piece of advice, you're young and still new to Iskar and the island.' She leaned in closer. 'And if you know what is good for you, do not speak of ghosts in this house.' Her expression was stern. But there was something more than a warning in her voice. There was fear.

I could not be silent. 'Mrs Lenister, why would it be better if I were not to talk of ghosts here? What is it that I feel all the time at Iskar? I can tell by your own face I am not alone.'

She stopped then and turned fully to me. Her pupils were black. 'Do you need to ask? We've all watched Mary, Miss Swansome. We've all seen it as I know you have. She will not speak to us but she does speak, doesn't she?'

My skin grew icy.

'Who do you think she is talking to?' she said.

I shivered, recalled the sound of the engine across the floorboards. *William*, Mary had written in the sand. *William.* I opened my mouth to speak but my tongue would not move.

'Miss Gillies witnesses it too. People say terrible things about Iskar and they whisper about Mary. We all guess who she thinks she speaks to, who she thinks she sees. Have you noticed the way her eyes sometimes wander to some space where there is nobody there and the recognition in them?'

I had.

'That language she uses when she talks, that was the way they talked to each other when he was alive. None of us could make it out even then. And why did they need to use one that none of us could understand?'

I felt sick.

'That is why you must not speak of ghosts, why we all stay silent. The doctor will be coming again in a month or so.' Her face grew serious and there was entreaty in her expression. 'Make her look to the living, Miss Swansome, and not to the dead.'

I remained at the top of the stairs watching her cap descend, my heart beating hard against my ribcage.

I sank to the hall chair. Beyond the window, the cold sea seemed uncrossable. Yet I could not imagine another day within Iskar's walls, another week. How would I endure all those things that were unbearable? I thought of Mary, with her kind thoughts but the actions that spoke too loudly of things that were unpalatable. I remembered home with a sort of impossible longing. Perhaps Alison MacAllister had been right. If I had stayed, at least I would have been with friends in a life that made sense. But then I thought of Mary again and my heart squeezed tight in pity and guilt.

Downstairs, the evening passed and I was aware of the crawl of the clock; how with each half-hour my discomfort grew. Too soon, I was climbing the stairs to my passageway where the salted air slipped in from beneath distant doors and travelled the corridors. The wind was in the eaves, and in the narrow passage to Hettie's room.

My thoughts were poisonous. I knew I would not sleep. The atmosphere chilled and flattened and shadows fell heavily along the wall. I opened the drawer for my book and set something rattling. When I leaned over to discover the cause, there lay a doll, face up.

I recoiled as though bitten. When my dismay was spent, I took it out. A particular discolouration of the torso made me suspect that it was the one that had been in William's room. I gazed down, horrified. The face that was no face. A scraped-away surface that was more repulsive for the absence of features. As I stared, a rush of knowledge came to me and I felt, in some part of me, that it watched me back. I slammed the drawer shut but the feeling did not diminish. My hand was unclean.

And all the time, I was aware that I was waiting, waiting for Greer's step on the corridor. And even though I expected it, it came, as always, as something newly insufferable.

I held my breath. My hands grew icy. Greer stopped outside my door and her purpose dragged at me like a piece of skin caught on a fish-hook. Her breathing curled on the air and her shoe fell still. I tried to resist her but my will bent to hers until, like the snapping of a line, she sighed and began her song once more.

I waited longer, wondering at the direction of her steps. They continued but then, instead of returning to the main hall, they were upon the garret stair. I listened. Halfway up they stopped and I felt her attention return to me.

My heart thumped. I went to the door, where I pressed my cheek to the wood. There was silence then a warm, sour breath came through the keyhole. A faint perfume. I sprang back with abhorrence. In that moment, all Greer's rudenesses ripped open my anger and I grabbed the handle and stepped out, but she had already gone. The passageway and attic stairs were empty.

Silence, and then a creak from the boards above. My hands were clutched tight as I looked up to Hettie's room. I took a step forward to challenge her but something in me faltered.

And then I was back in my chamber, pulling my cases from the press and laying clothes at the bottom; a new life beckoning somewhere else, one with a cheerful household and kindly mistress, children whose worst crimes were revealed in the jam-stained mouths of pantry raids or the leaping out from behind curtains to give their nanny a scare. I could see myself bent over a stream, collecting spawn for the schoolroom, and nature walks, pockets of conkers, acorns and beech nuts to roast over the fire, and while I packed, I told myself that Mary was better off with me gone. Miss Gillies would get another nanny, someone better equipped than myself. I had suffered too much to bear the weight of Mary's grief.

I dreamed of Edinburgh, the sound of hooves on cobbles, noise and crowds. I did not need to remain here.

I had made no promises. But by the time I had finished my face was wet with tears of betrayal.

20

The next day brought frost and the sky was blue with tufts of white cloud. Having made my decision to leave, I saw anew the island's beauty – the way light made the colours of Skelthsea surreal – and felt sadness that it was not to be.

Guilt ate at me that, like Hettie, I planned to leave without warning. I could not meet Mary's face in case she would read how my heart had turned. I imagined her, lost amongst the dusty rooms, with secrets and disfavour souring the air, shrinking further out of reach, further from well-being and hope.

I could feel the vibration of Mary's emotions as we returned from the sands. From time to time, she looked at me, a furrow of anxiety on her brow. She walked so close to me that irritation bit into my mood and I snapped, telling her to run off along the beach by herself. I could not bear to examine how I felt or to dwell on the way guilt twisted inside me. *Clara would not have run*, a voice whispered; *Clara was brave*. I gritted my teeth against the rawness of feeling until once again I felt only numbness.

The day passed in a blur, and after church Mrs Argylle came back and took dinner with Miss Gillies while Mary and I picked at ours in the playroom. The night drew in and I went to stand at her window where a triangle of ocean caught the moon. The vastness of the sea pulled at me like a chain. I tried to estimate how many days might be left before the boat next came. It could not come soon enough. After I had put Mary to bed I went downstairs to beg a hot drink from the kitchen and something to help me sleep.

As I passed the drawing room I heard Miss Gillies' voice. 'It cannot go on. She's half raving in these nightmares and they are not less frequent but more so, Greer tells me.'

'You've done your best.' It was Bridget. 'Perhaps Elspeth should be allowed to try a little longer.'

'The doctor felt she should have improved greatly by now and I fear nothing has changed. I must consider what is best for her. Mary is clearly not right. Not at all.'

'I am sorry. Those children were born under unlucky stars.'

'I think,' Miss Gillies said, 'that she must go. It's too late now for her to leave on the next boat but I will send a letter with it back to the mainland. The winter is drawing in and travel becomes more uncertain.'

Whatever Bridget answered I did not hear.

I placed a hand on the mantel to steady myself, knocking the peat scuttle. 'Is that you, Greer?' Miss Gillies called. I did not answer but fled upstairs. My hands trembled. I knew what Miss Gillies intended. She meant for Mary to go to an asylum. What of the three months? Had she given up already? I imagined the doctor coming up the path with his silver-topped cane and Mary's little case packed for the return journey to the mainland. I saw her standing on the boat watching the only place and people that she knew grow distant. And any affection that she had held for me would be a new wound upon her breast.

I stood in the cold of my room where my cases lay half-packed on top of the press. I had hated Hettie for abandoning Mary without warning. What did this say of my own loyalty?

I sat down hard on the bed, my chest tight with emotion.

What would Clara have said at my cowardice – she who cared so much for every creature? And I recalled the day a stray dog had wandered into the garden of Swan House and attacked our beloved spaniel, Bertie, as he lay dreaming in the sun. The sound had torn through the air as we ran to find the cause and discovered the shocking violence of a fight. 'Get a stick,' I had screamed to Barbara, because the flashing teeth and the frenzied hunger in the dog's eye left me too afraid to go near them.

And then Clara was there, shouting at the hounds and beating the stray with her fists and feet until Barbara came out with a pail of water and threw it at the snarling creatures. The stray fled, leaving Bertie shaking drops from his coat. And after we had checked that he was not hurt and changed Clara out of her wet gown, I had regarded her with wonder and shame.

Shame because I saw in her a courage I did not possess.

Candlelight caught my pillow and my heart stilled – there on the sheets was the bangle I had given Mary, and a note. 'I will miss you.' She knew. She had found my heart out before I had spoken of what was there.

A choke came up from my lungs. I touched the cold enamel of the bracelet – Clara. Clara who would never have fled from Mary as I planned to do. Clara, who I had left only for an hour or two because the dizzy scent of bluebells was in the air and I had promised to buy material for new dresses. I closed my eyes in remembrance.

'Go,' Clara had said.

'But I can't leave you alone.'

'Go, go, get ribbon too.' Her words, though slightly garbled, were clearer to me than the best orator.

And so I had gone, without even a kiss, snatching

my purse carelessly from the table and tying my boots in a flutter of sudden joy. I had secured a few more students. We could survive and I was filled with the pride and relief that knowledge bestowed. I remembered now how I had turned at the gate, Clara at the window, her thin, white arm waving wildly, and I had laughed. And the sun had washed across the grass and painted it brighter for my happiness. My beautiful Clara.

In the mirror, my lids were swollen and I witnessed my own grief for the first time, not as me but as Clara might have done, and some of the bitterness left me. I knew in that moment that I could not surrender Mary to the destiny of an asylum; that if I did, what remained of my life would be forever haunted by it. The room was still and the rightness of my decision left a deep sense that my destiny had always lain here, here at Iskar. To leave now would be a betrayal not only of her but of myself.

Calm washed my tears free and I took up the doll and the bangle and walked to Mary's room. She was not asleep but sat up against the headboard in a huddle of misery. Her eyes were hopeful when she saw me. I sat beside her. Taking her arm from the covers, I slid the bracelet over it and handed her Bobbity.

'I'm sorry,' I said. 'Forgive me.' And she buried her face in my chest.

'I promise you,' I said, 'I promise I won't leave.' And instead of returning to my bed, I climbed in beside her and held her warm body, not only for her comfort but for mine.

A few days later the boat came and the boat left and it went with Miss Gillies' letter and my resolve to stay. I unpacked my clothes and felt the tightening of Iskar's embrace. The sky darkened. Rain splashed upon stone and the sea battled with the rocks, spitting white spray against the cliffs. The boat was gone. I had made my choice.

This was my life now.

For better or for worse.

PART TWO

Night spilled from th'garret stair
And weaved a shadow on the air
Now something lives inside the spaces
Of those silent, secret places.

Anon (1748)

21

The day after the boat's departure brought bad weather: wind tore at the sea and threw surf high up on the strand. Nobody was surprised when the church bell rang out announcing that the service would be held that morning.

Muffled in our coats, we stepped into the blustery air and spray. As we approached the bay, the frenzied activity of gulls called our attention to where a group of fishermen stood around something on the pebbles.

'What is it?' I asked. Already the villagers were making their way down.

'Probably a big fish of some sort, a basking shark,' Mrs Lenister said. 'The birds soon find carrion, if it's to be had, and a storm like this will often bring it in.'

But whatever lay there was not a fish. As I moved closer I could see it was bundled in a greyish material, and I was gripped by a sense of foreboding. The pace of the household ahead slowed as we reached the parcel of sailcloth.

'It was trapped on rocks out at Gannet Point.' The

fishermen stood around in shuffling awkwardness. Then the smell hit us and I recoiled.

'Dear Lord,' Miss Gillies said. 'What in heaven's name is it?'

Mary's grip tightened on mine.

Nobody said anything. In the quiet, the sea rushed at the shore and a blast of wind tugged at the package. My eyes found a split in the fabric and something beyond, something white and curled like fingers of coral. Not coral, fingers of bone. I whipped round and saw from the shock on her face that Mrs Lenister had seen it too.

Robert Argylle came forward holding a scarf to his nose. 'Where did you find it? What is it?' But it was clear that he, like everyone else who now stood around, knew that the thing upon the beach was a body.

'It looks like a sea burial. The body must have come loose from its weights and drifted in the storm,' the fisherman said.

Argylle looked around at the pale, gawping faces as if assessing his congregation for absentees. 'A passenger on one of the bigger steamships?' he asked.

'That would be my guess.'

The mutter of talk began rising up to fill the hush, and I was aware of Mary beside me and turned to her. Her eyes were too big, too troubled, and snatched a heartbeat of horror from me. I moved a little away and pressed her face to my coat. 'Don't look.'

Another fierce gust caught the bundle and loosed a flank of hair that lay twisted in the sand like the wetted tail of a fox. I knew its colour.

Before they could be stopped, two black-backed gulls swooped in and tugged at the sailcloth, and the binding gaped. A piece of material was revealed. The colour was indeterminate, but there, glinting in the weak sun, was a brooch, large and oval, the gold casing dulled by age. My eyes sought Bridget Argylle. Had she seen it too, the cameo piece that had been so precious to its owner? My gaze found her, a little way up the path, a hand to her mouth. Mary's fingers gripped mine but I was barely aware. Surely not, I told myself, but I knew it was true. I knew who it was who lay there in her wrap of wet fabric. I knew that she was bones already, that whenever death had taken her it was long before I had imagined her in some city elsewhere, her palm in the grasp of another innocent child.

Silence fell and the fishermen's eyes slid to Miss Gillies. They too had guessed.

Mrs Lenister flicked me a glance with something like fear.

'We think . . .' The fisherman exchanged uncomfortable looks with his fellows, 'We think it might be yon nanny.'

'Hettie?' Miss Gillies' voice was tight with distress.

The fisherman leaned over and pulled something from the body then moved towards her with it on his outstretched hand.

'Mr Aird thought he recognized this, Miss Gillies.'

The brooch was passed to her.

Argylle crunched across the silty beach. He gave Miss Gillies a pitying look. I wanted to turn from the remains but I could not stop looking, revulsion looping my insides.

'To be found so close to shore, she must have died before the boat had got far on its crossing to America,' Argylle said.

The fisherman nodded. 'Those big ships trade with all the islands before returning. It's hard to say when she died and a storm can take something like this and carry it a long way.'

'She was not ill before she left?' Argylle asked.

Miss Gillies shook her head. 'She mentioned nothing in her letter.'

'Those big boats spread disease quicker than you can say.'

And so the questions and speculations began, and Argylle crouched over the corpse in prayer. The scar on Miss Gillies' face was livid against the blanching of her pallor.

I glanced up. Paterson was with his wife and

children. For once, the arrogance was absent from his expression. I searched the planes of his face, looking for anything behind it to hint at a deeper grief, but could not find it. It was then that I noticed the knowing way his wife regarded him and I believed that the truth was there.

Robert Argylle cleared his throat. 'We shall carry her up to the boathouse and then I suggest we give her prayer for such a premature end.'

The whispering rose – a bee-swarm of noise.

'I wonder what she died of?' someone said, starting a volley of questions.

Mrs Lenister went over and put a hand on her mistress's shoulder. I took Mary's fingers in mine and we made our way from the foul stench and up to the church. As we walked, I listened to the villagers talking of 'a doomed girl' and shivered, imagining her body cold, listing on the seabed at the mercy of the tide. Their eyes flicked constantly to our party and in their looks there was a sort of accusation.

As I followed, I could not take it in, could not imagine her dead. As we passed Iskar the windows of the attic were dark and pressing on my retreating back.

In church, the walls rang with prayer, and although Argylle's sermon was impassioned, there were no tears. As we left the pew, Greer gave me a stony look but there was a spark of colour on her cheeks.

The villagers lingered outside even though the rain began to snap upon the church tiles. My skin was numb and I hardly acknowledged Bridget as we left, although she clasped my hands in both hers, concern for me in her face. I did not feel the cold on my cheeks or pay attention to the dull ache in my ears.

Back at the house, the rooms were smoky, fire back-draughting from the chimneys. Mrs Lenister served coffee in the drawing room to which Miss Gillies added whisky and we sat for a while in appalled silence, interrupted only by the ring of china and the soft scrape as Mary turned cards in a game of Patience.

My thoughts strayed to the stretch of sand where Hettie's body had been brought in. The stuffed birds in their cages and the butterflies pinned to their velvet backdrop reminded me only of death.

'Mary,' Miss Gillies said eventually, 'why don't you run off and see if Mrs Lenister has some cake and perhaps bring down one of your jigsaws from the playroom to do by the fire.'

As soon as Mary had gone, Miss Gillies leaned towards me, her skin taut. 'I cannot believe that she is dead.' Tears hovered at the corner of her eyes. 'Dear Lord, when will it stop?' And she put her hands to her face. I got up and sat on the arm of the chair and placed a tentative palm on her shoulder.

'You've had a terrible year,' I said.

'Terrible. Terrible. How am I supposed to endure it?' Her expression was full of bewilderment. 'My sister, William and now Hettie. It's as if those that come close to me are cursed.'

I was aware that her feelings were all for herself and thought with pity of Mary and what she had just seen. However wicked Hettie had been, she had had the care of Mary for four years and Mary would feel it badly. How must she have felt knowing whose bones she was witnessing? Greer came in to collect the tray and I felt the high vibration of her energy as if Hettie's body had excited her. Rain began to patter on the window and candles flickered in the draught.

'Why is Mary taking such a long time?' Miss Gillies said, rising to her feet. 'She's usually such an obedient girl.'

'I will look.' And I left the room, taking the stairs slowly, mulling over events. The landing was dim and the whistling of wind in the eaves echoed through the passageway. Mary's door was open but her room empty.

The sound of crying drifted from above. I swallowed and steeled myself to climb to the garret. Outside Hettie's room I listened. It was quiet now.

'Mary?' I said with a courage I did not feel. 'Mary?' And I turned the handle.

She was curled on the bed, her knees drawn up nearly to her chin. I sat beside her.

'You knew Hettie very well,' I said.

She turned her eyes to me and covered them with her palms.

'You can cry, Mary,' I said, but she lay rigid and the room grew still around us.

'It's getting cold,' I said, 'and Miss Gillies is asking after you.'

She sat up, swinging her feet to the floor, and I rose also. Then in the mirror I saw her slip her hand beneath the pillow, take something out and place it in her apron.

Downstairs, the atmosphere was dense and with the exception of Greer, we all wore our shock upon our faces. Mrs Lenister served up a lunch that nobody ate. Lessons were cancelled and we remained in the drawing room. Although the wind grew in volume, the house was as quiet as a coffin.

I took out my sewing box and helped Mary with Bobbity's dress, but her stitching was careless and our hands soon fell idle. I was grateful when it was time for Mary's bed. Afterwards, I returned downstairs with an excuse for an early night on my lips, but Miss Gillies had placed a glass of whisky by my seat.

'We should raise a glass for Hettie, whatever I may have thought of her.'

I agreed and put the tumbler to my mouth. 'How old was she? She did not look more than twenty-five in the photograph you gave me.'

'She was twenty-four or thereabouts. A similar age to you. I must write to her family and send my condolences, although it's possible that the ship's captain has already undertaken that duty. Either way, there will be no boat quick enough to bring them for the funeral.'

'Not an easy correspondence.'

'No.' She put down her glass. 'Seeing that brooch again today reminded me of what an odd creature she was. She had a gift for finding lost things, you know – a ring, a piece of clothing or a key. Once she found an apron that Mrs Lenister had somehow packed away by accident in a mending pile.'

She cupped her glass. 'There was this one occasion when she found a necklace. I wracked my brains to understand how she had done it. Still, to this day, I cannot fathom how it was achieved.'

I looked at her questioningly. 'But lost things are so often fallen upon. Is it so extraordinary?'

'No, to give her her due, she had a talent. On the occasion I particularly recall, Greer had gone down to the pier. When she returned her chain was gone. It was of no value but meant a lot to her. Greer lost her family when she was young, you know.'

'All of them?' I experienced a stab of sympathy.

'All of them but her grandmother who was not – not a suitable guardian. The chain had been her mother's. Greer was most distressed. Hettie vowed that she could find it.'

'Where was it?'

'It had dropped to the beach on the stones there. Almost impossible to see, but Hettie went straight to it apparently.'

I leaned forward with interest. 'Had she been out with Greer? Perhaps she saw it fall.'

'No,' Miss Gillies shook her head, 'that is what is so mysterious. She had not been out at all that day. She was with the children all morning as rain kept them in.'

'Greer must have been very grateful,' I said.

She gave me a curious look. 'I expect that she was.'

'You say that her grandmother was unsuitable?' I said.

She paused and frowned, seemed about to say something but thought the better of it.

'I must write to Hettie's family now, Elspeth. Onerous as the task will be. Goodnight.' Her gaze lingered on my face with a strange expression and then she left the room.

As I ascended the stairs, a new anxiety attacked me. I stopped at Mary's door and remembered how I had seen her slip something into her pocket earlier. I was fairly certain what I had witnessed. Careful not to wake her, I searched her bedding until I found it. It was heavy in my hand, the hair rough against my skin. This pebble I had not seen before. There was no doubt as to the hair's colour – even in the dull light of the candle it burned red. A binding spell. One that bound a person to someone they loved or to someone they wished to die. My heart sank at Mary's disobedience. I had not thought she would play further with them after my admonition but I could not doubt her intentions now, not at the moment I had made the decision to stay.

I closed my eyes, saw Hettie's body unfurling from the canvas like strange coral, her hair ribbons of red kelp. I gazed down at Mary's sleeping face and wondered what thoughts lay beyond the pale lids.

Wind tore at the house and my heart banged. I held the stone warm against my skin and tried to sense if it gave off poison or desire, whether it spoke of love or hate.

But my palm was blind.

22

The day of the funeral brought dark skies and dense cloud that scudded low over the sea. We set out to a damp wind that pulled at the black overwear we wore. The screech of gulls was a mockery and Miss Gillies' steps hesitated as we reached the valley. One or two others had left their cottages to join us but most stood and watched in silence as our procession ascended to the tolling of the bell.

The church was full of shadow. Hettie's plain coffin stood at the front, catching the spill of light. There were so few of us that we filled only two pews. The seat creaked behind me and there was Ailsa, who regarded me with an expression I could not read. Nobody spoke until Robert Argylle climbed to the lectern and began his prayer. There were not enough voices to give it body and the words sounded as thin as an afterthought. There was no hymn, and soon, six men arrived in heavy coats and the coffin was carried out of the doors and on to the ridge.

Unlike tradition on the mainland, women were allowed to watch the interment and we followed behind. The crofters had not moved from their houses but remained huddled to observe our sombre journey.

The smell of wet earth was in the air and my eyes slid to the dark cavity that had been prepared. Argylle led another prayer and then Hettie's coffin was lowered into its waiting place. A handful of dirt was thrown and it was over almost as quickly as it had started. We left to the sound of spades and the shovelling of earth rattling on to the lid of the box.

Ailsa stood beside me. 'Doesn't it make you wonder what purpose the binding spell was woven for?'

I did not answer. I could not answer. To believe that such spells could work was too terrifying to comprehend, but the memory of what I had so recently found beneath Mary's pillow was heavy on my thoughts.

'This is a terrible business.' Bridget had a spot of pink on her cheeks. 'It must be uncomfortable for you, Elspeth. Hettie was your predecessor after all. I can see how this has hit you.' She put a hand to my arm and squeezed gently. 'Come visit as soon as you are able. It has been too long since you last came and I can see you are anxious.'

I smiled but my eyes went to the Fiaclach where mist coiled around the stones. A falcon circled above and screeched into the wind, bringing sweat to my skin. The thickened sound of soil on soil told us that the last sods were being laid. The men bent over their work, coats lifting in the wind, and Argylle stood hunched at the graveside, his face cut with sadness.

We said goodbye to each other in hushed tones. Miss Gillies pulled at her gloves and we began the descent to Iskar. I longed for night, for the privacy of my room, but the clock ate away at the time with aching slowness. When it finally struck nine, I climbed the stairs to the corridor and even the shadows and the presence of Hettie's empty chamber above me did not cause my steps to falter. I shut the door behind me and sank on to the chair by the fire.

I had not known that tears had been so close, but once alone, I wept until my body was shaken empty. I cried for Clara. I cried for Mary, and finally, I cried for myself, for my lost home and my lost loved ones and last of all for my helpless, crippling unhappiness.

The funeral left a stamp of desolation on all the inhabitants of Iskar and a colder tone breathed its dampness into corridors and corners: in rooms with heat, the walls sweated condensation. The air became sour and even when the rain ceased, the creaks and groans of the house were punctuated by a constant drip of water from broken gutters.

Mrs Lenister twice dropped a piece of cutlery and once knocked a milk jug across the tablecloth. Her dark eyes were brighter and her cheeks sunk beneath shadow. Greer continued unaffected and I sensed a superiority, a knowingness about her that drew my suspicions. I saw no grief, no sense of loss. Had she shown Hettie her hostility – the same hostility she gave me? I took to watching her more closely and examining the antipathy between us. At times she would stop in her work and challenge me with a stare. I had ceased trying to placate her and met her dislike with an open show of my own. But the person most affected was Mary. For someone who took up so little emotional and physical space, who made so few demands, she diminished further and as hard as I tried to cheer her, I watched her shrink.

My promise to stay was the candle I held to the darkness. I walked less and often spent my free hours waiting for lessons to finish. I spent hours cajoling Mary to speak. The doctor would already be in possession of Miss Gillies' letter, be planning his return trip, might have informed the asylum of the patient they might soon have. If I had been afraid and unhappy before, it was worse now.

It had been a long day and I had found it hard to fill the hours usefully. Even though I dreaded Greer's lullaby, I was grateful for bed. I lay between the sheets and waited for her to come as I did now each night. The pale flame that burned on my candle did not touch the depth of fear. Opening the drawer, I took more, placing them on every surface, and then I waited and eventually it came. I did not move but listened, and as she neared I resisted the tug of her voice, although the vibration of it wove into the air like the pluck of a cello string. I put my hands to my ears to block out the sound but still it continued on my brain. When I took my hands away, she had gone.

In spite of barbed thoughts, sleep dragged me down.

I had been dreaming, running through the streets of Edinburgh, my dress aflame, and yet the fire did not warm me and then I was on the ridge, with Skelthsea below, a ruin of ashes. Only the headstones in the graveyard jutted from the scorched earth. A crow, huddled on Hettie's plot, regarded me with a preternatural gaze and opened its beak.

It was not song that fell from its throat but a whistle that soured and sickened on the air and wound like a shadow over the island.

I was awake then with the heaviness of dread across my chest. A dream, I told myself, but then I heard it, that thin note which shrank the marrow in my bones. Not the wind, my mind whispered, something else.

I jerked to sitting, dazed momentarily by the multiple yellow hearts of the candles. The room was empty although the flames twitched, sending shapes sprawling to the walls.

But the room was not as it had been when I had taken to my bed. Upon the dresser, someone had placed two figures. Leaning against my brush set, they appeared to be standing: dolls, such as the ones I had already found. I stared unbelievingly. One, I recognized as the object that had been in my drawer, but now it was dressed, its jointed arms flat to the side, red hair glued to its scalp and a face painted upon it. Next to that was another – its head chalked white and in boys' garments. They were grotesque – repugnant. My thoughts were still half in my dreams and I did not want to get up to investigate, did not want to take a step further into the reality of what I was witnessing. With the darkness behind them they could have been standing unaided and the waking part of me began to pulse with horror.

I shrank beneath the covers and then the female figure, blown by wind or too uncertainly balanced, fell forward and toppled to the floor. I jolted and, as distinctly as if it was beside me, there was a sigh.

I leapt up and searched but found no culprit, no Mary or Greer, hidden in my room to play a night-time trick.

I forced myself to lay them upon the bed. Just dolls, I said, but part of my mind had become detached from rationale and I was back in a place of numb distortion. The faces stared back at me, two sets of inanimate eyes, but in spite of their vacancy I felt observed and I locked them in my dresser with disgust.

There was in the air a sentience that had not been there earlier. My eyes fixed on the drawer; even closed, the dolls' presence was diffused in the room and I could not quell the rising panic. I sat for a minute and tried to contain my trepidation, then I stepped into the corridor to see if someone lingered, but it was hushed and hung with the emptiness of early morning. So I returned to my room and lay watching the flickering shadows and listening to the sound of sea and wind, and in spite of all my resolve not to be afraid, fear cut deeper into my bones.

23

It was a few days until I was able to leave Iskar in search of the relief of my friend's company.

'I thought that you would come. Hettie's body turning up like that is a dreadful thing. Poor Miss Gillies must feel it terribly,' Bridget squeezed my hand, 'but I think it's worse for you.'

I was grateful for her sympathy. I opened my mouth to mention the dolls but held back. Would she have laughed? She would have seen my discomfort immediately and there was a little voice that told me she might think it had been Mary who placed them; I did not want Bridget to think ill of her.

'The house has been sombre since we learned what happened to Hettie,' I said.

'Unchristian as it sounds, don't you think it's something of a blessing that she will not be able to ruin another child? But Mary . . .' She paused. 'I had wondered if such a dramatic thing might cause her to speak.'

'She has not.' My eyes roved the cheerful walls, the rise and fall of the dog's ribs by the fire. How much easier, I thought, would my job have been had

I nannied a child in surroundings such as these, in a place where nothing unaccountable could happen, a place I could not imagine steeped in shadows and where dolls could not disarm me.

'She will not be missed,' Bridget said.

'I hear there was at least one who might,' I ventured.

She put her head to one side in question.

'Paterson claims to have had a relationship of some sort with her. Is it true or a boast?'

She laughed. 'Him. That is true, but how deep his feelings ran, who can say? He is a womanizer to the marrow. It was very well known. I was surprised when Evangeline kept Hettie on after the affair was so much discussed. Such a bad influence.'

I thought about Paterson and Hettie. They made a pretty coupling.

I'm assuming he's already tried his hardest to tempt you, Elspeth?'

'He has. I do not like him at all.'

'I suggest you stay well away. He is not the sort of man who has respect for the boundaries women such as us would impose. He has had many an affair here under the nose of his wife. He will take whatever he can.'

'Was his relationship with Hettie a long one?'

She sighed. 'I believe so. It was said that she was very heart-struck with him.'

'And yet she left to go to America?'

'Yes, perhaps it was over then, or she had bigger fish to fry.'

'Paterson was her last lover here? You said she had taken a few.'

She nodded. 'As far as I know. And how is Iskar? Is Greer still a thorn in your side?'

I laughed dryly. 'A little, yes. I have given up trying to win her good favour; her manners are certainly amiss. Miss Gillies mentioned to me that Greer's grandmother had been an unsuitable guardian.'

'Did she indeed?' She chuckled. 'True enough. Greer's grandmother was little regarded. She was a cruel woman, but more than that she had some very ancient beliefs that made the islanders fear her. She was said to be able to curse.'

'Are you telling me that she too was a witch?' I heard again the sound of Greer's step outside my room, the swell of her voice and then the hush, and shuddered.

'I suppose, yes, if you believe that sort of thing.'

It was Greer that I pictured then, not Hettie, winding hair about the pebbles and whispering spells. I thought of Mary in the same house with unease.

The wind had dropped as I returned to Iskar, leaving a strange stillness. In the hall, I shook out my hat and as I passed by the parlour, the door was a little open and Miss Gillies' voice came out from beyond. It wasn't unusual for lessons to finish early or for Mary to be left alone with a piece of learning. Miss Gillies was standing with her back to me at a far window, Greer beside her. Miss Gillies said something which Greer answered and Miss Gillies patted her hand. I stepped back, shocked once again at their intimacy. Miss Gillies talked in a low voice.

'There, Greer,' she said, 'don't worry, my dear. You can trust me.' I had never heard such a tender tone before but as she spoke, Greer's eyes slid to the reflection and to where I stood. I moved away but not before she had seen me and made the tiniest adjustment to her look and brandished her triumph.

The rest of the day, I tried to hold on to the peace I had found with Bridget, but slowly the fear seeped back like an incoming tide. I thought of the dolls lying in my drawer, their sightless eyes open to the darkness. Much later, as I made my way to bed, I stopped and let myself into Mary's room. Her expression was peaceful in the glow of the candle, but as I sat beside her, I felt the shimmer of misgivings and the queasy knowledge that I did not truly know what lay behind her face.

I lit candles but I did not find much comfort in their thin light. From the table, I took a book and tried to ignore the sounds of the house around me, tried not to focus on Greer's nightly visit. I wondered now what ran through her head as she stood outside my door but could not afford to let my thoughts travel in that direction.

Restless, I got up and went to the window and stared out on to the island. It was in full darkness now with a slice of moon veiled behind cloud. I laid my head against the glass, feeling its cold touch to my skin, and as my eyes grew accustomed to the lack of light, I saw something in the Fiaclach near Stack Mor. I leaned in to the image, searching for what had caught my attention, and there it was – someone had lit a lamp. It was not possible to see the face, but the light caught the movement of arms as they worked at the patch of ground and I knew by the dread that crushed my ribs that it had not stopped with Hettie's leaving or William's death – that they had not taken their practices to the grave. Someone here continued to whisper spells and spill blood in the name of some ungodly religion.

I thought further: what if Hettie had not been the one to bring the dark art with her? What if she had been taught by someone here – Ailsa or Greer? How many of the islanders still clung to such primitive beliefs? And felt once again that I had been tricked.

I was woken by a thin scream that brought me to jolting wakefulness. Mary. Fumbling into my shawl and picking up a candle, I rushed to the corridor. It was silent now but for the hissing of the oil lights. The door to Mary's room lay open. But not only hers, the door to William's wing was also ajar. There was a bang from beyond, and another cry.

I had no choice but to enter, the flame casting trembling shadows to the walls. I paused in the corridor where faint noises came from above. Passing William's room, I came to the stairs and looked up; a sound came again – a whimper. Clutching the candle, I ran up the steps and to an attic – home to broken furniture, stacks of trunks and two large presses.

Mary sat in the centre with her back to me rocking on her knees. I sped forward and cast the light over her stricken features.

'Mary,' I whispered but she showed no sign that she was aware of me. 'Mary,' I tugged at her shoulder and placed my shawl about her. 'We must get you back to bed.' But then, as if someone was making an effort not to be heard, there was the distinctive timbre of a foot upon the staircase behind.

My heart began to bang and I stood tall although my legs felt boneless. 'Who is it?' My voice cracked and there, with her unhurried movements, loomed Greer.

Her eyes went from me to Mary and narrowed. 'You should have called for help.' Her voice was low and full of distrust. And in that moment I recognized that she truly scared me.

'Did you open the wing?' she said, coming closer. 'I know it's not the first time that you've been to see these rooms.' Her eyes darted to the furthest corner of the attic as if she might find evidence of what we had been doing there. I swivelled to see what she was looking at but was presented only with more disorder. When I turned again her look blazed with fury.

'I have no interest in this place, Greer. Mary had a nightmare. I found her here.' And was chagrined by the fact that I had explained myself when there was no need.

'Mary,' I said, putting a hand on her arm, 'come, it is time for bed,' and I took her hand and pulled her to me.

Greer's gaze raked my body. 'Give me the key,' she hissed.

Mary's feet were glued to the boards and her lips faintly blue. 'Get a hot drink and heat the pan for Mary's bed immediately. She will catch her death.'

'The key.'

Anger and frustration rose in my chest. 'That will do, Greer. I do not have it. Now go, before I wake Miss Gillies whose only concern will be for the welfare of her niece.'

'If you say so,' she said, so quietly that I nearly did not hear it, and left.

Mary had not moved; her scrutiny remained fixed

to the window where the panes were black but for the dusting of a moon.

I was then assailed with a strange discomfort and the room swam in something dark. 'Mary.' My voice was a little louder but I followed her gaze. It was not the window itself she was fixed on but something half hidden in the rafters, something barely distinguishable from its surroundings. When I finally made sense of what I saw, I shuddered. A breeze seeped in from the frame and it swung a little in the draft, eliciting the faintest of notes. A widows' whistle. My skin crept with disgust and I turned to Mary, but instead of her fugue-like trance, her face had thawed. She looked me full in the eye and gripped my wrist so hard that I nearly dropped the candle. She was struggling with something.

'What is it? Tell me. What is it?'

She opened her mouth and I waited, but then her lips clamped together, forcing the silence. Through the quiet another creak sounded on the floorboards. Mary too had heard it and her gaze swung to the landing.

I walked Mary to the stair top and cast the candle to the treads below. The steps fell ahead, a slight turn halfway down with a rail on one side – smooth as glass. And then, through the silence, the widows' whistle began to whine and a shape emerged, one that seemed cut out of something blacker than the night itself. It could not be, it could not, and yet there it was, but even as I stared into the dark heart of it, my feet frozen to the boards, it had gone, leaving on the air the press of its unnaturalness. Behind me, the widows' whistle fell to silence.

It was Mary who took the steps first, breaking the spell and leading us to the landing where I followed on trembling legs. We walked the passageway through the smudged darkness, my breath quick in my lungs. I fixed my eyes ahead on the goal of the exit, swallowing the horror of what I had seen and felt.

In Mary's room, I lit candles and got her into bed, where her shivering continued. I wrapped my arms about her, aware that as much as I gave comfort, I needed it for myself. I recalled the moment she had been about to speak, felt the urgency of her desire to communicate. My hands shook. From the table I picked up a notebook and pencil.

'What is it you want to say? Trust me,' I said. 'Trust me.'

Her eyes blinked wider with fear, but she took the pencil and with a trembling hand wrote:

I am going to die.

24

I slept more deeply than expected, coming awake to a sense of unreality. A dream had taken me to Gulls Cry, where I had stood in front of a mirror but it was not my face I saw there – it was Hettie's.

Mary lay beside me. I had not had the courage to return to my empty room and its shadows. The moment on the stairs played through my head again like a nightmare and Mary's note still lay clutched in my fist.

I did not want to believe Iskar haunted. In the bright air of the beach later, what had passed the night before seemed dreamlike – I told myself that I had not seen what I thought I had. That some trick of the light had caused that shape to appear on the treads. I told myself that the widows' whistle had not begun to whine. But even as we walked a part of me was whispering. *You lie*, it said. *You lie.*

Mary did not run to explore ahead but remained beside me, her eyes uninterested. I took her hand. Mary's conviction that she would die. I tried to reject it, but it struck every time, chiming with some deep conviction of my own.

And the widows' whistle: if there were ghosts, was this how they came? Could it really be possible that there existed an instrument made by human hands that could exert such supernatural power? I dismissed the idea, told myself that it could not be so, but in my head its thin whistling played through my bones like a sick motif.

In the hall, I helped Mary off with her coat. A fire guttered in the grate, exuding the sweet scent of peat, and although it was barely eleven, it was as though night were already falling.

'Why don't you go up and change and I will sort something hot for you to drink.'

As always, she obeyed without question, climbing the stairs neither rushing nor dawdling, and I watched her retreat uneasily.

Lunch came but I could barely taste what I put in my mouth.

Miss Gillies broke into my thoughts: 'Elspeth, I have a favour to ask of you.' I looked up.

'I have been going through some of Evangeline's correspondence and realize that I must follow up a letter I received from an Edinburgh school in relation to William's education. I really thought I had collected any post of importance from Gulls Cry but it seems not. Although we were not close, I do not feel particularly comfortable being amongst her things and I wondered if you would be kind enough to see if you can find any more letters relating to that in her drawing room desk.'

'Of course,' I said.

After lunch, I took the key and made my way back outside. I walked slowly, recalling the last time I had stood in those frigid rooms in search of my charge.

Taking courage, I unlocked the door and walked inside, pushing from my mind the sound of that other step I had heard before. In the hall, I was met only with silence.

I made my way to the drawing room and the elegant desk of walnut. A pen and inkstand lay in dust beside a sheaf of notepaper. I sat in the chair and opened compartments and pulled out drawers.

I soon found the bundle of letters and began to go through them. Most were personal but I came finally to a thick envelope embossed with the Edinburgh school of which Miss Gillies had spoken. I emptied the contents to check that the correspondence was there and came across another letter that instantly caught my attention.

Dear Evangeline,

As I reach the end of my life, I have come to regret some of the decisions that I have made. It saddens me that it has taken me too many years to come to this understanding. What occurred between yourself and Violet is long in the past and I do not want to go into the next life without making amends.

I made the decision some years ago, after what happened between yourself and your sister, to change my will and testament in favour of Violet. I now believe it unlikely that Violet will bear an heir and I have changed my will again, as follows: Iskar and the estate shall be passed to William when he comes of age, and in the event of his death, to Mary. It will be William's responsibility to care for his mother. Violet is to inherit Gulls Cry and be given sufficient monies to ensure her comfortable living. The full details are clearly stated in the will and testament.

Violet has been appraised of these changes. It is my greatest desire now that you and your sister will also leave the past behind and use what time you have left to make friends of each other.

Ernest William Patrick Gillies

Shock pinned me to the chair. So William had been due to inherit Iskar and with his death it passed to Mary.

It was impossible to imagine Miss Gillies at Gulls Cry and the decay was evidence of that. I did not believe that Miss Gillies intended her father's wishes to be fulfilled. Of course, if Mary were confined to an asylum from which she did not return, the question became redundant. I thought about the shine of antipathy Miss Gillies could not hide when she regarded her niece. With Evangeline dead, was Miss Gillies the only witness to this change of will?

I collected the correspondence from the school and returned the letter concerning the change of will to the other papers. Mice moved behind the wainscotting, scratching and scuttling, and I was keen to be gone. With William dead and Mary in an asylum, only Miss Gillies would be left to rule her beloved Iskar and I recalled her words that day in the morning room: *I love it as well as I might a husband.*

25

I arrived at Iskar to find that the hall was cold. No fire burned. The smell of something unappetizing drifted from the kitchen. I returned the key. Miss Gillies was in the drawing room in her customary seat, her head fallen back in sleep. I studied her with new distrust, then backed out, closing the door quietly.

There was still some time before tea and I lay down upon my bed. When it was four, I made my way downstairs where the chilled hallway was speckled with sand. There was no smell of coffee or food.

The house felt different, apparently empty of its inhabitants, and I stood to listen but there were no noises suggesting activity in the nearby places.

Finally I reached the chapel and stood at its entrance. I heard Miss Gillies then and put my hand on the knob, but her voice came, hard and angry.

'Tell me, Mary. May God forgive you. Tell me the truth. Why don't you speak?' There was silence, then the groan of a pew. I pulled my palm away and felt the urge to turn and run.

Miss Gillies spoke again, her voice cutting through the corridor like the snap of a whip. I wanted to protect Mary from the verbal blows but cowardice sent me fleeing back along the passageways and to the drawing room, where the flames had nearly gone out in the grate. I crouched, ashamed, and built the fire up and waited. It was some twenty or so minutes before they joined me. Miss Gillies' face was drawn and pale, her lips colourless. I searched Mary for a response but she looked as she so often did – as closed as a locked room.

'Mrs Lenister is unwell having succumbed to the head cold that is going round,' Miss Gillies said. 'Greer too. Angus has gone to fetch Mrs Brodie who sometimes does for us.' She closed the door and leaned in close. 'Her cooking is nothing to Mrs Lenister's, but we must smile and pretend gratitude.'

With a sigh, she levered herself into the chair and pulled her shawl closer.

'Mary, go upstairs and wash. When Mrs Brodie arrives, she will be able to prepare our tea.' Miss Gillies dusted her hands. 'The chapel is filthy.' She glanced at me. 'Mary and I went to say a few prayers for those so recently lost.'

There was a noise in the hall and then a knock and a woman I recognized from the island entered, her face lined with smiles. She was as small and hunched as a beetle, with leathered skin and thin lips.

'Mrs Brodie,' Miss Gillies rose with a look of relief and accompanied the woman to the kitchen.

The evening continued in an atmosphere of unease, not helped by the tasteless stew and hard scones prepared by Mrs Brodie. The smell of oil hung in air already smoky from peat fires. I played a game listlessly with Mary, but exhaustion made me yawn often. Mary went to bed and I came back downstairs.

'We shall miss Mrs Lenister and Greer,' Miss Gillies said. 'It is when they are not here that you realize how much they do and with so little fuss.'

But all I could think of was how much cleaner the atmosphere was without Greer's influence. Until then I had not fully grasped how oppressive her presence had become to me.

'Miss Gillies,' I said, 'something has been bothering me.'

She looked up in surprise.

'Greer is not as courteous as I would have expected.'

Miss Gillies sighed. 'Don't mind Greer. She wanted your job.' She sipped at her coffee and grimaced, putting it down. 'Awful coffee. I expect she holds some jealousy for you and your position here, but you must be used to servants; surely Greer's manner does not concern you?'

'Not a great deal,' I lied. 'I wondered if she took out her jealousies on Hettie as well, whether that was part of the reason Hettie left as she did.'

'You're not thinking of leaving, are you?'

'No, I intend to stay.'

'You have that wrong, Elspeth.' She paused. 'Greer and Hettie were friends. Best of friends.'

I had not expected that. I had assumed that, like me, Hettie was disliked. There was no definite reason, I realized now, to have made such an assumption.

'Had they been friends since Hettie came to Skelthsea?' I asked.

She shrugged. 'I suppose they might have been. It was only when Hettie lived here that their intimacy was so apparent.'

The idea unnerved me but made strange sense. It made me more suspicious that Greer shared Hettie's beliefs. And if that were not the case, and Greer did not practise witchcraft, how could a friendship be justified between one who had gone so bad and one who had not? And how could Miss Gillies remain so undisturbed by such knowledge?

Miss Gillies stood then glanced in the mirror and her jaw tightened as if she was witnessing her scar for the very first time. From the decanter, she poured two large whiskies. I noticed the worn patches on her dress and loose threads at her wrists.

'Did you ever think of living away from Skelthsea?' I asked.

She was reflective. 'I'm too old for change now and the estate would be hard to sell. I have lived here all my life and have not left since I was seventeen.' She put a finger to her cheek. 'People here know my blemish so well that it has become invisible. I went once to Edinburgh afterwards and . . .' She paused. 'It was like being burned once more. The looks. The pity. Never again, I told myself, and so it will be. This is my home and always will be now.' Her face was slack and her cheeks unnaturally red.

I wondered at her comment regarding Iskar. It was not, if I understood the newer will and testament to be true, hers to sell.

'Do you miss Edinburgh?' she asked.

'I miss aspects of it, yes. Others, no.'

She smiled. 'Always the diplomat, Elspeth. I sense you have not found this job easy so far.'

I blushed. 'It's had its challenges.'

'You miss your sister a great deal?'

My chest tightened with the sudden pain of it. 'Always.'

She was quiet.

'Do you ever miss Evangeline?' I dared.

She did not seem to mind my question. 'Not as much as I should.'

'That is very sad. Did your personalities not work well together?'

She was reflective. 'No, it was not that. She did things that I could not forgive her for. That is the truth of it. I am not one of those carefree people who finds it easy to forget the wrongs done to them.' She lifted the glass again to her lips.

I wondered what it was that could have been responsible for such a breach in their relationship, but to ask would have seemed an impertinence.

'And when Mary is independent, where is she to live?'

'Gulls Cry is hers.'

More than once she topped up her glass until her eyes glazed. The room grew cold and the fire sank in its grate. I thought she had fallen asleep and I opened my mouth to say that I would go when she turned, instantly alert.

'Iskar was left to me,' she said. 'It has always been mine and always will be.' Her voice was as unforgiving as flint.

And as I sat there, her lies settled upon me and fell like cards upon falling cards, their faces hidden, overlaid by others. I no longer knew where the deceit ended and reality began. My thoughts burrowed past Miss

Gillies' skin, searching for something I could believe in. But, finding nothing, I could only stand and wish her goodnight. There was nobody I could trust, I realized. If Mary were to remain on Skelthsea, I alone could achieve it.

26

The following morning brought frost and a breakfast of overcooked egg with more weak coffee. Miss Gillies eyed me over her cup with a sigh.

'How are Mrs Lenister and Greer?' I asked.

'Mrs Lenister is improving but Greer still has a heavy cold and has lost her voice. I don't think we can expect her back at work for a few days yet. I must see if Magda can manage extra hours. There is so much to do. More than Mrs Brodie can manage.'

The day was fine and Mary and I took the kite to the beach. The sun lit up the valley and clouds ran in the sky as white as goose feathers. Mary unwound the kite string and as the triangle of red took to the wind, her lips parted in something like a smile and for a moment she was as any other nine-year-old, without a care.

I wondered at the things inside her head to which I did not have access. But I smiled too as I looked up, watching as the kite was whipped towards the blue. It swooped and climbed and her hair lashed about her cheeks, and for the first time, I saw a likeness to her mother and aunt. One day she would be beautiful, and I tried to imagine her growing into womanhood, but again my mind refused to create the image. I could not visualize her anything other than this young girl, and my fear for her fell like a shadow over my thoughts.

Later we took out our sewing boxes to finish Bobbity's dress. The gown was finally complete and Mary buttoned the smock and held the doll up to admire her new appearance.

'We shall have to make her a whole new wardrobe with dresses for all occasions.' I fished in my pocket. 'Now, Bobbity,' I said, 'I have a small gift for you.'

Her eyes were questioning.

'This is for Bobbity's dress.'

My excitement was almost as great as hers, and with grave ceremony, I passed her a pin with a ladybird enamelled at its centre. It was a trinket, picked up at some market and of no monetary value. But for a moment, Mary went completely still as if it were made of gold and studded with jewels. Then, carefully, she took it and fixed it on the doll with fingers that trembled. I looked down at her face and my heart squeezed. After it was done, she looked at me and in an impulse flung her arms about my neck and placed a kiss on my cheek.

I laughed at her delight and returned it with one of my own.

Mrs Brodie had prepared a lunch of hard biscuits and a tasteless broth which we ate in silent distaste, and then there were lessons and I left Mary to her aunt.

I was putting things away in the playroom when Mrs Brodie came up to re-stock the fire buckets. She had a quick, nervous manner but despite her slight frame she was strong, making short work of it.

'And how are you enjoying Skelthsea, miss?' she asked. 'A bit different from the city?'

'It is,' I admitted, 'but I like the work.'

'Well, that's good now. Poor Mary. We all felt for her losing her brother like that.' She had a coarse, high voice that reminded me of the gulls. 'And,' her tone was tentative, 'do you believe that she's getting better?'

'Yes,' I lied. 'She's improving all the time.'

Mrs Brodie's eyes were flat with doubt. 'And Iskar, miss. What do you think? It's a fine house, wouldn't you say?' She laid down her brush and I could feel her keen anticipation for an answer.

'It is a fine house,' I said. 'And you will be filling in for Mrs Lenister until she's better?'

Mrs Brodie nodded. 'I will. I'll be here again before dawn, but I go back to my own home at night.' Her look was pierced with some intent.

'I expect you have your own family to get back to?'

Her eyes darted to the door. 'Not young ones, no,' she paused. 'I could work later and stay overnight if I chose.' There was something in her tone that made me look up.

She watched me conspiratorially, a sheen of excitement on her face. Her voice dropped, 'You see, miss, not many of us would stay here the night, you know. Only Mrs Lenister, Greer and yourself. I bet it's why yon Hettie left.'

'What do you mean, Mrs Brodie?' I was irked at her forwardness and because she had agitated my own doubts when for a moment they had retreated.

She leaned closer. 'They say there are ghosts here. That Mary talks to ghosts.'

'There are no such things as ghosts,' I said, although my mind echoed back that I no longer wholly believed that.

'He has been seen,' she whispered.

'Who?' I said but my heart already knew. Hadn't I thought I'd glimpsed him too – upon the beach and in reflection in Mary's bedroom?

'William is seen from the attic rooms at night looking out.'

'You cannot see into a window from outside. It's too dark,' I said with more force than I intended, upset to have my own fears affirmed.

'Not when someone lights a candle. People have

seen him. A candle on the sill and his face at the glass.'

There was a tremble along my fingers and I put them on my lap to hide it and recalled the sound of his engine on the boards. 'It must be Mary. She goes up there from time to time and they were very alike.'

But her look told me that she did not believe me. 'He's seen late at night, at the window, just a candle on the sill.' She smiled coldly. 'He was up to mischief in life and he can only be up to mischief in death.'

After she had gone I sat and tried to conquer my feelings. I could sense that she had tried to unnerve me and I had observed her greedy enjoyment of my discomfort. Or had I misread her and she had told me only to warn me? And I was cold to my bones.

The silence was too heavy. I knew that even if that were not the truth, that there was some truth in what she had said. The idea that while I slept, William stood in Hettie's room above to stare out at the island was too dreadful to digest. I must not allow myself to think of it.

But if there were ghosts they were passive, surely, beyond will and thought, only leaving their imprint upon the earth? They could not touch or harm, and yet that was not true either because did not Mary talk to William?

I thought of Ailsa and her knowledge of the

pebbles and the widows' whistle. If ghosts existed perhaps these things too had a power I had never before acknowledged. I remembered the pipe in the attics, recalling its shape and for what it was hewn. Surely not, I told myself, but before I knew it, I was hurrying out of the house and up the valley, reaching Ailsa's out of breath.

The croft stood in near darkness; only the thinnest twist of smoke spiralled on the grey clouds. I knocked but it took a while for her to answer, and one look at her flushed skin and bloodshot eyes made me ashamed of my visit. She too was ill and I turned to leave, but she took my cuff and pulled me in.

For once she sat with her hands idle.

'You are unwell. I should go.'

'You are here now and I can see that what you want to say is important.'

The kitchen smelt sour and I regretted my decision. 'You have told me of the widows' whistle, of the pebbles and what they were made for. Do you believe that such objects can achieve their purpose?'

She grimaced. 'Why do you ask?'

'You know why I ask. You know.' In spite of myself my eyes pricked and I wiped angrily at them with the back of my hand. 'All is *not* well at Iskar,' I said. 'You know this. It is full of things that . . . things that do not make sense in the world I understand.'

She coughed into a cloth and when she had finished a sweat stood out on her brow, but I could not continue: it was as if in uttering my fears aloud they would become more true.

'Listen to me, Miss Swansome.' Her eyes seemed to reach so deep inside me that I crossed my arms across my belly. 'This is not about truth. What is truth after all? What single thing can you pluck from the universe and hold up as reality? For that you would have to trust your eyes, your ears, your senses. People will say that those who believe in magic are mad, but whose judgement do we believe? You are here for one reason only: because you no longer trust your own.'

'I don't trust my own,' I agreed. 'All the beliefs I have spent my life in certainty of no longer seem to tell me the whole story. I see things, hear things, begin to know things that I never thought possible. Never.'

She was utterly still. 'Then you have your answer.'

Just before I left, I turned.

'Thank you, Ailsa. Be well soon.'

The wind was rising, crushing peat smoke low over the roofs and ripping at the sea, and I felt as if I had been hit by some sort of madness.

That night, climbing the stairs, my feet started to drag and I scolded myself. At the top, I paused and anchored myself to the newel post. The passageway stretched ahead, swallowed in unsteady light. At the end, the blackness was solid. Once again, the faintest of whistles came like a blade of wind along the eaves. I thought of Miss Gillies and the servants far away in other wings and tried not to let it cower me. I almost wished Greer was well enough to pass with her lullaby.

Frost coated the window, distorting the moon, and above, the weight of Hettie's room pressed down upon me. I lit the fire and three candles and then stood by the hearth. It was quiet then but for scratchings and bangings from the room above. Rats and birds, I told myself. I ignored them and sat at the dressing table to brush out my hair, but just as I was ready to climb beneath the covers, steps sounded on the garret staircase. I sat rigid. And there was Greer with her lullaby. But Greer was ill. She had lost her voice. Yet I could hear her – as clear toned and pitch perfect as ever. Had Greer's illness been another lie? Greer's or Miss Gillies'?

From some reserve, I summoned the last of my courage. And with the foolish optimism of a gambler placing his final coins on a low hand, I got up from the stool and snatched at the door handle. I did not give myself time to think before stepping out. With crippling dismay, I discovered that the corridor was empty, even though the final note still rang in the half-silence.

I looked desperately both ways, gazed up to the garret and to the gloom of the main staircase. It had to be Greer. The creak of a foot sounded again on the attic treads and I moved closer and stared upwards. I yearned then to see Greer, but the stairwell remained empty. Then slowly from the coiling shadows a shape began to form, clotted in the blackness, and from its darkness something began to emerge. I could not take my gaze from it nor quell the way it gripped my fears in a vice. I wanted to shout, to call, but I was trapped by the horror of it.

'Stop it,' I managed to gasp at last, and the feet stilled. There was a shift in atmosphere and with slow deliberation, I felt the other consciousness turn to me and rest its eye upon mine. I could neither move nor scream.

I tried to draw away when something crept from its mass and pressed itself to my skin, the palest of touches, and with it, something else twined and bloomed in the air – a pitiless rage and hatred that left me breathless.

I stumbled back. The cadence of the hummed lullaby began again and wove around me, and its step began once more, up and up until it reached Hettie's door. I was unable to unglue my feet because finally I knew: it had never been Greer who had walked the corridor at night. It had never been Greer who sang at my door. It had been Hettie. Always Hettie, from the very first night. And taking a key, I locked my room from the inside and lay on the covers, unable to stop the violent shaking while some note of stale perfume still hung in the air about me.

27

As I dressed, the shivering truth found a home in the very bones of me. Skelthsea was unbearable. Buttoning Mary's dress, I caught my reflection in the mirror – my eyes wide as though still receiving the horror of Hettie's night-time visit. My fingers were clumsy, and twice I had to undo and re-braid Mary's hair. It seemed to me that I was more girl than woman as if, at some point, I had tricked myself into believing in my own adulthood.

After breakfast, wrapping her warmly, I took Mary and we walked towards the north beaches. Idly, we picked over shell and stone although neither of us was enthusiastic. It was when we reached the woods that I beckoned her and we climbed to a sheltered spot between the pines. Our feet sifted the scent of needles. Laying down my scarf upon a trunk, I took her hands in mine.

'I need to talk to you, Mary,' I said. 'I know that you can speak. You must use your voice now.'

Her eyes recoiled.

'You have no need to hide the truth from me. I know that you see them – William and Hettie.'

She pressed her lips together and watched me nervously as if trying to read my thoughts. I took her hands tighter in mine, 'I have seen them too. You're not mad.'

Both relief and pain broke over her features.

'Hettie walks the corridor by my room nearly every night. Do you hear her? We have to talk about this if you want me to help you. You know that, don't you?'

Her skin, already pallid, whitened further. 'I know you speak to William. Did he tell you to make the pebbles and the dolls? Was that him, Mary? Did you dress the little dolls and leave them for me?' Because I believed then that the dolls might hold their own potency; that they, like the pebbles, were wrapped up in the ability to draw ghosts. My voice cracked. This thought was so distasteful that I could hardly bear it.

At last she shook her head. I reached into my bag and brought out paper and pencil. She began to tremble but she unclasped her hand from mine.

'Tell me about Hettie and William,' I said. Wind shivered in the trees and the first drops of rain began to patter on the boughs above.

She placed the tip to the paper but pressed too hard, going through the thinness, and it was as if a little wound had opened there.

'Hettie,' I urged and she lifted her head, turning to where Iskar lay beyond the rocks.

Hand bunched over the paper, she began to write.

'Hettie is dead but not dead.'

'Why does she come back?'

She bent again. 'I do not know.'

'And William?' I whispered.

'He comes for the truth.' Her face darkened.

'What is the truth, Mary? What is it? If you know it, tell me. You can trust me. You know you can trust me, don't you?'

She looked down again, her arm trembling. 'I cannot say,' she wrote, and she closed her lips in a stubborn line.

'You cannot say or you do not know?' A gull's screech ripped the air. 'It's too late for secrets. We must face the truth together to make this better.' But she shook her head so violently and with such distress that I relented.

'And Hettie?' I could barely drag the words from my tongue, 'Does she speak to you?'

Her expression slammed shut. I could not bear the thought that she might visit Mary in the night and whisper in her ear.

'How can we stop them?' I said. 'How do we make them go away?' And in that moment I was more child than she.

She gave me a fierce, sad look that I could not

interpret. 'How can I help you, Mary?' But my heart was not seeking help for her but for myself; it was I that could not endure it.

She put her fingers in mine and leaned against me. After a pause, she scribbled again.

'If you leave, will you take me with you?' and my heart squeezed as tight as a fist inside my chest.

'Listen, Mary – William and Hettie were bad. They did bad things. They are still bad. I believe that they make you do bad things now. There is one answer only. If Hettie talks to you – you must not engage and you must stop talking to William. You do talk to him, don't you?'

She nodded. The fear was back.

'It must cease today. Promise me, Mary, promise me.'

At church, although I knelt and clasped my hands together and joined in the song, my head was empty but for the fear of the coming night. Mary stayed close beside me as if she longed for the touch of another living soul.

Robert Argylle read out his passage but its words had no meaning. Instead I studied the dark corners and the hovering spaces in the apex of the roof. Not even the coughs and creaks that emanated from the congregation made a mark on my discomfort.

And then it was a supper of overcooked meat and undercooked potato. Greer reappeared from her sick bed and watched me maliciously, a pearl of pleasure in her iris. And I thought then, that however odious had been the idea that it was she who had stopped by my door each night as she hummed, how much worse it was that it was not her at all.

The lamps behind Miss Gillies dropped light on to the table and across her face. I had ceased to notice the scar by then – it was as if my brain had readjusted her features, making a whole from the undamaged cheek. But that night, the opposite was true, and more than once, as I answered a question or made a comment it seemed to me that her whole face was twisted.

I was not aware of my fork as it lifted food from the plate or my hand on the knife. Miss Gillies smiled and I watched the way she drank her wine and how every now and then, her gaze slid to Mary and displeasure come down upon her expression like a veil. *Iskar is mine*, she had said. Time dragged and yet passed too quickly, and with each turning of the hour fear became a little brighter on my skin like a burn. Finally, Miss Gillies rang the bell and announced that she was retiring for the night.

And as I climbed the stairs, the house was grotesque and fear began to whine inside me like an insect. I trusted nobody. From the drawer, I took more candles and placed them about the room. I could not face the dark and the things it hid. Already, I was shivering in anticipation of Hettie's lullaby, but I wanted it to be over, wanted to hear her come and to hear her leave so that I could at least not have the agony of waiting.

So I sat in the chair, skin singing with fear, and I waited until the cold stung my hands and face and night turned to early morning. And still I waited, and even as I sat and heard only silence, I felt her outside my door, breathing quietly as she stood there, a gleam of malice in her dead eye, her intent as dark and malicious as anything I could envisage. But she never made herself heard. And in some way, it was not better. It was far worse.

28

I barely felt the cold as I dressed. It was the cold inside that occupied me, and the realization of how horrifying my life had become. Outside, Mrs Lenister stood at the jetty with her basket and my heart lifted.

'It's an icy one.' She shivered dramatically in her coat, giving us a smile that creased on her inflamed cheeks and ended with a cough.

'Are you well now, Mrs Lenister?' And I was so relieved to see her face again that I could have wept with it.

'Can't complain. It'll be good to be back. I was cared for at my sister's, but her family do not have room for me there in all truth. Ah, let me get some fresh fish. You like a fish, Mary. And you, miss?' The eyes that searched mine were concerned. 'Are you well, Miss Swansome?'

'I am well,' I lied.

Mary wandered off, her braid and coat tails whipping up in the wind.

'She does not seem better, poor thing,' Mrs Lenister said, giving me a pitying look. 'But I must get my catch. And we can't trust Mrs Brodie to cook them properly, can we?'

I shook my head and gave a smile, but she did not smile back and there was in her eyes something beyond pleasure, a spark of concern.

When lessons began, I found my feet on the path to Bridget, but at the street I paused. What would I say about what I now believed? I knew Bridget would find it silly and suddenly I was making my way to Ailsa's again.

I did not like her look of triumph. I took my place on the settle.

'So soon?' she said.

'The dead,' I said. 'They can return, can't they? They can walk outside a room or turn on a corridor. They can speak.'

'I thought you did not believe.' I clutched my hands in my lap. 'Do they speak to you, Miss Swansome? Are you telling me that the dead talk to you?'

I swallowed. What had I come for? Denial? What would denial do now? Would it change what walked at Iskar? Would it stop the visitations to Mary? But I would not tell of that. The gossip did not need more fuel.

My heart hardened and I gazed her full in the eye. 'Hettie is not in her grave. Some part of her is at Iskar.'

Ailsa took her pipe and twisted tobacco into its

bowl. I could feel the strength of her thoughts. 'Is it only to you she comes?'

So she believed. There was no doubt. I felt then as if my bones must crumble to dust.

Mary. In all of this she was the most vulnerable. I imagined Hettie at Mary's bedside whispering a lullaby in her ear. Whispering other things – how to scrape the face from a doll, how to dress it up again and paint eyes upon its head. I shuddered. I did not want Ailsa to think that Mary was sullied.

'I see no sign that others are haunted.'

She smiled. She did not believe me and I looked away. Ailsa knew, as Mrs Brodie knew, as Miss Gillies and the whole island knew – that Mary talked to ghosts.

'There is something else,' I said. 'There are dolls. Ones such as a child might have but they have been disfigured – have been remodelled to look like a specific person.'

Her gaze was intense.

'They are not natural.'

'Who do they depict?'

'Hettie and William.'

The room fell to the breathing rush of the fire and the sea. Air grew tight in my lungs.

'Remember what I told you of the binding spell? Well, the dolls too; they are made in the image of life for the same purpose. Whoever created these wanted to tether Hettie or William to Iskar.'

But they are dead, I thought. Who could have wanted such a thing? I wanted to be sick.

'Who do you think made them?' she asked.

But I could no longer reason and I stood shakily and went to the door.

'I told you once before,' she said before I left, 'and I will tell you again. You must take care.'

Greer was polishing the floor when I returned. She looked up briefly, her dark hair curled in sweat at the nape of her neck, and a frisson of open dislike passed between us. I did not care. It was too late to court her goodwill. Exhausted from the previous night, I slept until lessons finished and found Mary in the kitchen baking a cake with Mrs Lenister. The thought of the night to come was already ticking in my blood, but I sank into the rich, sweet smell of the room and helped Mary to cut fruit and stir the mixture in the bowls.

Mrs Lenister chatted as she worked, imparting island gossip, and for a while it was as if the rest of Iskar did not exist. Here, there was only the warm bubble of Mrs Lenister's domestic ordinariness and Mary dipping a finger into the bowl to lick the sweetness from it, and I could have been anywhere.

Evening came too soon, and in the drawing room the clock ticked and there was the occasional snap of Miss Gillies' scissors. As bedtime drew ever nearer, the night came turning its key upon my fear and all I could hear was the wind on the ridge and how it howled across the graves, graves that had not closed the eyes of those that lay there.

'You're very pale, Elspeth,' Miss Gillies said. 'I do hope that you have not gone down with what ailed Mrs Lenister and Greer.'

'I am well,' I insisted, and returned to the book that could not engage me.

'Is Skelthsea not suiting you?' She paused. 'Or Mary?'

'Mary is well,' I said. 'I mean, I enjoy my position.'

'Something else?' She put her work in her lap.

I paused. 'Iskar can be a strange place,' I said tentatively. 'So many noises and creakings. I hear whistling and steps when there is nobody there. I have heard it has a reputation for being haunted.' I do not know what I expected her to say. Would I have taken comfort, even if she had confessed that she too had experienced ghosts?

'This does not sound like you. The island likes to gossip and if it can't find something to say about the living, it will conjure up something to say about the dead. You don't believe it, do you?'

'No, no,' I insisted, but my answer was late coming

and an odd expression crossed her face. She rang the bell and Greer brought whisky.

'You know,' she said as she took a sip. 'It is a strange house.' Her eyes scanned the room reflectively and I could see that her thoughts had wandered from where we were and were roaming the corridors and rooms unused for decades, rooms that now collected only dust and secrets.

'I was remembering Iskar today – Iskar as it once was,' she said finally. 'I was thinking about Evangeline too. We were close as children, or closer. I was listening to you and Mary playing hide-and-seek recently and it reminded me of a game that Evangeline and I played. We called it, "Whistle and I'll Come".' Somewhere beyond, an owl's screech tore at the night.

'It was a little like hide-and-seek.' Her hands twisted the tumbler. 'Evangeline would always hide first – an elder sister's privilege – and I used to wait on the red-backed chair in the hall. If you look you will see the loose thread I used to pick at when I grew bored. I would shut my eyes tight.' She blinked hard. 'Sometimes, sitting there, I would hear nothing but the creak of the boards on the landing and the sound of servants far off. I would wait and wait and then I had to go up the stairs and listen. When the time was up, Evangeline had to whistle. My uncle taught us, although my parents forbade it as unseemly.' Her eyes flashed. 'I think that added to its appeal. I would start at the top of the house and listen so hard until I heard her. The rules were that we would whistle twice, every five minutes or so, until we were found.' She looked at me curiously. 'Did you and Clara play together as children?'

I was back instantly with the hot square of sun that came through the glass in the window seat of the parlour. And the weight of her slight frame as she leaned against me while I read from some adventure book.

I nodded.

'Evangeline was easy to find. I was always the cleverer.' Her chin lifted. 'One day, her whistle took me hither and thither. Strictly speaking, the rules of the game forbade us moving from spot to spot. I was so cross. It led me finally to the old servants' wing, which was out of bounds. We were not really supposed to go into their quarters, but during the day they were empty and so I went in. The whistle took me up to the attic where rooms were used as storage.' She gazed into the distance. 'I remember clambering between the dusty boxes and crates to stand at the window. The windows are tiny up there, no bigger than a handkerchief.' She finished her drink and her hand strayed to the decanter again. 'I remember looking out and wondering what lay beyond the sea and believing that one day, I should find myself living in some grand house in Edinburgh with a family of my own.' Her fingers went unconsciously to her cheek. 'Of course that wasn't to be.'

'Was your sister there?' I asked.

She turned to me as if she had forgotten that I was present. 'I heard the whistle again and rustling, something moving between the boxes. Of course, I knew it was Evangeline. I was angry by then. She had broken the rules. "Just come out," I said, "or I'll tell Mama that you came up here."' She paused again. 'But it was the strangest thing, Elspeth – the whistle came again but it was not my sister's. It sounded – it sounded like a flute or a pipe.'

I went cold.

'There I was stamping round the attic, dust flying up to sting my eyes and spoil my frock.' Her gaze fixed on mine now. 'I was filled with the most terrible dread.' Her hand went to her neck. 'And then I heard Evangeline upon the stairs and she appeared at the door.'

I was stunned.

'I thought afterwards that it must have been a servant or a guest – we had so many – but that feeling, it stayed. Not long after, well . . . there was a . . .' She looked at me, doubt hovering in her expression. 'Not long after the servants talked of a ghost walking that part of the house.'

She knew. She too knew that the dead could return.

'Are you saying you thought what they said was true? Iskar was haunted?'

She hesitated, playing with the bracelet on her wrist, then looked at me slyly. 'I don't believe in ghosts, Elspeth. There are no such things.' But behind her look there was a shadow of fear.

'Whose ghost was it said to be?'

'It was very sad. Tragic. I'm surprised that you have not heard already. The island loves a good tale. One of the servants, a scullery maid, fell from a window. It was most unfortunate. It was her they said had returned.'

The room grew still, the hairs on my neck pricked awake.

'Accident? Murder?'

'Your thoughts are too ghoulish. It was an accident.'

She leaned over and topped up both our glasses. I did not refuse. I welcomed anything to numb the creeping dread of night. The fire cracked and a handful of rain hit the panes, sending out a thin rattle. The whisky worked its way hotly down my throat.

'Who was the servant?' I asked finally.

Her voice grew sober. 'It was Greer's mother,' she said.

Greer's mother? Greer with her solid antipathy – the way she moved as if heavy with bad feeling. It made sense, however, of the strange loyalty and affection that Miss Gillies showed her. How much guilt and responsibility might Miss Gillies feel if Greer's mother had died within Iskar's walls?

Miss Gillies leaned towards me, her pupils huge and black. 'I thought I heard that whistle again recently. But see, I am letting my imagination walk too freely, as the servants did. It will be just one more loose tile among a thousand.' But as she rose her face was ashen. 'And now you must excuse me. It's been an exhausting day.'

29

Each step I took to my room was laboured, and once I was in my bed, even with so many candles, fear kept tapping at me. I thought of Greer's mother dying at Iskar. The sound of the whistle. I could no longer deny it. A widows' whistle to summon the dead – to draw souls from their graves. Miss Gillies, hearing the whistle in the servants' attic, and not long after, the ghost of Greer's mother being witnessed. And had I not heard that very sound? Had I not heard it on occasions when Hettie came? Perhaps the dead needed to be called from more than the heart – maybe they could only be pulled from the grave with magic.

I sat up, shivering in my nightgown. Hettie and William had not just returned at a whim – no, they had been summoned. Something in me shattered then. It was too terrible a reasoning to contain. And who had been the one to call them? Was it Mary? Or was it Greer? She had been Hettie's best friend. Had she called her back because she loved her? I thought of that presence and the feelings it sucked from me – and knew that no comfort could be found in such a creature, and I was thankful that Clara had not sought me from death. If Hettie had not been called for love, then for what? I knew now that Greer's grandmother had older beliefs. Did she teach them to her granddaughter? And if she did, and it was Greer who called them, for what purpose? But even as I asked myself this, I knew in my heart it was to do ill.

I recalled the rage that blistered the air when Hettie's ghost had stood before me, embroidered from the very darkness itself. This was beyond anything that I could ever have imagined – too far beyond wicked.

I must have fallen asleep because I woke from a dream. I had been back at Swan House. At first, it had been as it once was – sunshine patterns on the walls, busy fires and comfortable, fraying sofas. I had walked from room to room, past pictures and tables, but at some point, the scene had changed and I was pacing not a furnished home but a blackened shell. The sunny walls dripped with soot and charcoaled furniture lay in ruinous heaps and I began to run. 'Clara,' I screamed, 'Clara.' And then I dreamed that I was in the corridor outside Hettie's bedroom.

There were no doors, just gaping holes. Weeping came from inside, soft but insistent. Clara stood at the gutted window looking down upon the bed where three dolls lay – Hettie, William and Mary. The dolls were as stone. Their blind painted eyes were closed.

Something shone on Hettie's abdomen, something that drew my attention. I leaned in, expecting to find a jewel, but as I grew closer it was to find a piece of torn and bloodied skin. As I watched, it squirmed against her dress and Hettie's white fingers came down and clasped it in an embrace. I leaned in even closer and her eyes flew open and saw me.

I was awake in an instant, gasping. The candles had died and the room was in darkness – I was not alone. There was a breath next to me, not soft but rasping, and the mattress fell away at my side. I turned with horror, the scent of rot in my nostrils. The scream in my throat did not come but I was out of bed and rattling at the door for escape. Once open, I ran into Mary's room.

All was quiet. Mary lay asleep. When my heart had stopped its frenzied beating, I took a candle and returned to gaze at my bed. Perhaps I had not come fully awake and had been suffering the tail end of the nightmare? But when I put my fingers to the sheet there was dampness and some smell of the sea. I began to cry then, not for grief or for pain but for my insufferable terror.

When I finally stopped, the eye of the moon was on my face and I realized something else. I knew that I had shut the curtains before sleep, yet now they were open.

I went and stood at the glass and gazed out at the place I had come to hate and saw someone coming down from the ridge with a bundle. I gazed and gazed until finally I recognized her from her walk and posture and the way that she held her head. It was Greer. She was out at night when everyone else was asleep. Greer. And in my mind's eye, I saw her lips upon the widows' whistle, summoning Hettie from the cold soil.

I could not tolerate my own bed and so I went to Mary's and lay beside her, and as I closed my eyes, I saw the dream again and the three dolls upon the bed – not just Hettie and William but Mary too, and fear lay on my heart like a stone.

The morning was sharp and unforgiving. I shivered as I dressed. I could barely look at the mattress. I recalled Greer coming down the valley with something in her arms. Whether she blew the whistle or not, I knew that she was here, in the weaving of the mystery, but her thread was hard to pull loose.

It was early and I was gripped with anger for all the things that Greer might have done. Dressing quickly, I crept downstairs where Mrs Lenister's tuneless humming and clattering drifted from the kitchens. Greer was in the library, dragging a hearth rug to beat out the dust. I stood and watched her. Her movements slowed. It was plain that she knew I was there but she took her time turning. Her expression did not change.

'I saw you,' I said. 'Last night. Coming down from the Fiaclach. What were you doing there?'

If my statement concerned her, she did not show it. She watched me carelessly, resting a hip provocatively against a table.

'You do not scare me, Greer. I know what you are. I know your wickedness.' I had not realized the violence of my anger until it was in my voice. 'You have done your worst and it hasn't worked. I am not leaving. Do you understand? I know you have things to hide and I will find them,' I hissed. 'So do not make more of an enemy of me.'

She neither moved nor spoke but at the corner of her lips I saw the beginning of a smile. With a huff of exasperation, I left.

Mary and I spent the morning by the sea; away from the house my unease lessened. I felt better for chastising Greer. Fear was the enemy I must master. If there were ghosts, I must face them. They did not harm, I told myself; it was the fear itself that would hurt me if I let it.

The day was bright and cold and the sea glittered like a field of coins. In the rock pools we found urchins and gobies and Mary hardly paused to check Bobbity in her pocket.

With a sudden impulse to be rid of my unhappiness, I bent down, undid my boots and removed my stockings. I placed my foot on the wet sand, felt its grain on the arc of my foot and the tiny bite of shells. I laughed and Mary did the same. Soon, we were racing across the beach, boots swinging from our fingers, and up to the tideline where I gave a tiny scream as the first wave washed over my ankles.

We returned to the house happier after our game.

It was during tea that I pleaded a headache and went to lie down. Later the banging of the door and voices drifting up disturbed me and I went to the window where lanterns and lights marked the procession to church.

Miss Gillies was on the path with Mary and behind her the servants. Angus's lantern swung and Greer's face was caught as she turned, the angle of her head cast back at Iskar and towards my window. My hands clenched at my sides and I knew what I wanted to do. What I must do.

Taking a candle, I took the corridor by the scullery at a run and then the flight of stairs that led to the servants' garret. I needed to see where Greer slept, see if there were things that would tell me more about her.

The herbs that Mrs Lenister used were in the air and I peeked into her room curiously. It was small with a simple bed, a press and dresser. On a table lay an open Bible and a photograph of two babies in knitted hats and jackets. Starched aprons hung from pegs.

I continued on to Greer's room. If she was involved in witchcraft, perhaps the evidence of it would be there. Greer's chamber was of similar proportion. Her drawers contained little: some bangles and brooches, a hairpin with missing jewels. I wondered if she had stolen them. There was even a brush and comb set although the silver had worn nearly away. I walked across the floorboards in case, like William, she had hidden something there, but it occurred to me that if she was as clever as she seemed to be, she might not keep anything that would hint at aberrant behaviour in her own room, and so I began to explore the remaining ones.

There was little to show but cobwebs. It was then that my thoughts returned to the night of Mary's nightmare, how her anxious glance behind me hinted at something to hide, and I raced back to the scullery for the key.

Once in the attic, my eyes went to the window where the widows' whistle had hung and was dismayed to find it gone.

I stood then exactly where Greer had stood that night and looked to where her gaze had rested. When

I went to investigate, I discovered behind some crates a Japanese screen, the painted flowers long faded.

Carefully, I pulled the screen away. The floor creaked uneasily beneath my shoes, bowed from age and damp. Wind threw splashes of sleet at the windows.

Behind the screen lay a large cupboard. I eased it open and held my candle up to see what lay beyond. The object filled nearly the entire space and was concealed beneath a piece of discoloured cloth, but I knew immediately by the shape what it hid and I was part intrigued, part horrified. Until that moment, I had not known what I might be searching for, but my fingers pricked with discovery. Crouching, I pulled the fabric away and placed the light closer.

Although ancient, the doll's house was beautifully constructed. Paint peeled on the roof and on the splintered walls. It was a portion of Iskar, its black windows caked in grime. I tugged at the hinged front and it opened to Iskar's drawing room with chaise longue and even a tiny cage of stuffed birds. I studied it, fascinated. The library, the parlour, the morning room were all there, as was a great kitchen with miniature copper pans hanging from racks. I did not have time to linger in wonderment. The house even had attics, and it was in one of these that I found the dolls – stiff and dusty and repulsive to me. I reached and plucked one out. Like the others, it was strangely warm and I swallowed a lump of disgust.

They were dressed in clothing of an earlier age but there could be no doubt that the ones that had appeared throughout the house came from the same family. But why was it hidden so? Surely this had belonged to Evangeline and Miss Gillies. And by the worn furnishings, it must have been greatly loved. Why had it not passed to Mary?

I rested on my haunches as the doubts and questions came creeping. Who now could be responsible for the dolls in my room but Greer? I imagined her thick fingers filing at the wood with slow enmity.

I leaned in further and there, beneath one of the beds, were two more dolls and I was seized with fear. There could be no doubt as to who they represented: one was myself and the other Mary.

With a gasp, I sat back. What should I do? But the answer was already in my head – I reached in and took them, then collected the rest lest Greer should fashion more.

It was as I rose from my knees that my eyes found it, laid on the floor of one of the rooms – the widows' whistle. I hesitated, but only for a moment, and then I took that too.

I hid my finds in various places in the spare rooms that lined my corridor. It would take hours for Greer to find each one, if she decided to search. All evening, I was aware of what I had done and one minute I experienced a strange elation and the next a crushing anxiety. When Greer next went to the doll's house to execute one of her schemes, she would find what was missing and guess that it was me.

That night, I lit candles and lay down, watching light flicker on the walls. Perhaps, at least, Hettie would not come. Had I not taken the widows' whistle? But as I lay there, my skin tight from fear and some exhilaration, her step and lullaby came anyway and I buried my face to the pillow, hand to my ears, until, when I took them away, there was silence once again.

30

Rain flashed at the windows and the sea rose and threw itself upon the rocks. I had dreamed that I was back at the MacAllisters' on a day when the sun burned through the glass of the panes and on to my cheek. I had been lying on the bed, my neck and hands wrapped in dressings. The medicine I had been given was wearing off and I was in that place of gnawing realization that Clara would never return. From the open window the perfect blue of sky and the scent of sweet pea had been unbearable and I was ripping at the dressings, pulling skin from the wounds; the sound of feet running on the stairs. I was unaware that I had been screaming. Some of the shock of that moment and the many others of those early days returned again, making my throat harden with pain.

After lunch, in spite of Mrs Lenister's concerned looks, I went out into the wet. I had to be away from Iskar.

A curl of smoke drifted from Bridget's chimney and I thought longingly of her company, but some compulsion to revisit the graveyard found me walking that way. I do not know what I expected to see – the earth cleared and the coffins standing empty? I knew that whatever dark magic made it possible for them to slip from death and haunt the living was achieved by more than the ability to free themselves from the physical confines of their plots. But I wanted to see for myself that the soil lay undisturbed, although it could bring little comfort.

Mist seeped in from the ocean and I walked slowly to Evangeline's grave and then to William's. Even away from the house I felt haunted. I knew that I should not be there with the ground still fresh from burial and I did not like the way the cold air wound about me, or the untethered screech of gulls, but somehow I was standing over Hettie's grave.

I shivered. It had grown colder and a sea-fret had closed in, falling across the valley and obscuring the view. I felt a moment of alarm and made my way to the gate.

It came then, braided upon the other sounds, and panic swept through me – Hettie's lullaby. I tripped, put my hand upon the mossy dome of a marker and recoiled at its fleshiness. My breath rasped in my throat. The cloud now was so dense that I could no longer see and I imagined her just behind me, watching me with her dead gaze, reaching out a hand to touch my face. I stumbled out of the gate.

Once I was in the valley, I no longer knew what

was up or what was down, and a new terror began to grow. How often had I been told to take a stick? If I sat and shuffled along, I should surely be able to make my way back to the safety of the valley without injury to anything other than my coat. A damp footfall sounded in the grass behind me and I whimpered, imagining Hettie in her dress of dead skin.

There was a breath at my ear and before I could react, the heel of a hand sent me tumbling down the hill. Although I covered my head with my hands, stones tore at my skin and clothes. It is too late. It is too late, I told myself, and I thought of Clara and Papa and part of me welcomed death.

I remembered no more.

I woke in my room to pain and Mrs Lenister's gentle hands cleaning my wounds. Miss Gillies paced up and down behind her.

'Elspeth.' And there was so much relief in her voice that I was momentarily gratified.

'How is your head?' Miss Gillies said. 'You were lucky.' Her face, pale in the afternoon light, was afraid. 'They found you in the valley. Goodness knows how long you had been there. Skelthsea is dangerous. The mists come down in an instant. Have you not been told often enough? You must learn to take a stick.' She was angry now.

‘I was pushed,’ I said. ‘Someone pushed me.’ But even as I said it, I remembered the sound of the lullaby and a renewed horror crept along my skin. A look passed between Miss Gillies and Mrs Lenister, one that told me they did not believe me, and I sank into the pillows and closed my eyes.

‘You’ll be fine, now. Don’t fret,’ Mrs Lenister said and adjusted the covers. ‘Nothing broken, just cuts and bruises. I expect you’ll be proper sore tomorrow.’

Miss Gillies left and Mrs Lenister pulled up a chair.

Her kindly expression made me want to confess everything I had so recently learned. Did she know that the doll’s house lay in the attics? Did she know about the widows’ whistle?

‘Miss Gillies told me about Greer’s mother,’ I said instead.

Mrs Lenister was quiet. ‘That was a long time ago.’

‘Not for Greer, though,’ I said. ‘For her, it must always feel new.’

‘It was a lot for a girl to take in, for sure. It would’ve been better had she not witnessed it.’

‘Greer saw her mother fall?’

Mrs Lenister flushed. ‘I thought Miss Gillies told you. Greer was in the gardens when her body hit the ground.’

‘But why was Greer in the gardens? She must have been too young to work here.’

'Greer was but six or seven, a wee girl then. Mr and Mrs Gillies let her come to work with her mother. Iskar is such a big place and Greer was a quiet, easy little thing in those days.'

I reached out and took her hand. 'Please, Mrs Lenister, please tell me. Something is amiss with Greer. She means Mary and myself harm. You told me once that you knew everything that went on in this house. You must know she's not right.'

A flicker of fear hovered in her eye but she did not reply.

'She's like Hettie, isn't she? Does she believe she's a witch?'

'Well, I cannot say that I know that to be true.'

'Can you account for her dislike of me? It is no common dislike – it is stronger than that.'

Mrs Lenister wiped a hair from her brow with the back of a forefinger. 'I thought at first it was envy of you: a young woman with education, a good background and nice clothes, but perhaps it is more than that,' her tone was cautious.

'She liked Hettie, didn't she?'

Mrs Lenister gave a nod. 'They were an odd pairing.'

All the further things I knew, felt too unsafe to say. I remembered Mrs Lenister's sharp retort when I had suggested ghosts. If I spoke my mind now, would she tell Miss Gillies? Miss Gillies might think I was as mad as Mary. And so I lay back against the pillow and closed my eyes again.

I started to drift and thought that Mrs Lenister must have gone when her voice lifted me from sleep. 'Morvern, her name was,' she was reflective. 'Greer's mother. She worked below stairs.' She paused. 'She wasn't . . . she wasn't quite right.'

'How was she not right?' I tried to sit up but the pain had me lie flat again.

'She was an unfortunate creature. A childhood accident had taken an eye and she was a nervous, quiet girl but also slow in her mind. There was a big staff in those days and, I'm sorry to say, they treated her very badly.'

Through the window, the scene was wintry. 'What did they do?'

'She was given the worst jobs and teased horribly. They gave her tasks that brought lots of scolding. She was very unhappy. Mrs Jeffrey, who was housekeeper then, did not keep them in order as she should have done.'

'Miss Gillies said it was an accident. Are you suggesting that she killed herself because she was unhappy?'

Mrs Lenister shrugged. 'Perhaps.'

'Did the Gillieses know about Morvern's treatment by the staff?'

'They won't have concerned themselves with below-stairs matters like that.' But there were things that she was not saying, I could see it in her eyes.

'What is it? Mrs Lenister?' I shifted up the sheets. 'What is it you're not telling me?'

She picked at the bedspread and lowered her voice. 'It wasn't just the staff. Evangeline and Miss Gillies took to teasing Morvern too, calling her hither and thither, giving her ridiculous tasks that were made to humiliate her. I don't think they really meant harm. They took their lead from the servants, but it was unkind.' She paused for a beat. 'Greer was not like her mother. Where Morvern was slow, Greer was sharp, even at six – she saw it all and she understood it.' Mrs Lenister put out a stubby finger and wiped a piece of soot from her sleeve. 'She had a bad time of it, miss. It won't hurt you to be kinder to her, to forgive Greer for some of her ways.'

I recalled the glimpses of intimacy that I had witnessed between Greer and Miss Gillies. Surely Greer would have detested a person who had contributed to her mother's death. 'So Greer must have hated them – hated the sisters.'

Mrs Lenister put her head to one side. 'It was complicated. Although they teased Morvern, they made rather a pet of Greer. The girls were nearly grown then at sixteen and seventeen.'

'So Greer did not resent them for their treatment of Morvern?'

'Well now.' She shifted on her seat. 'I did not say that. Greer is a complicated creature. She spent much of her time up in the nursery where they had their toys. She was small for her age and had a quietness about her, and often I would come across her without having noticed she was there, if you get my meaning.' She frowned. There was a pause and I felt her pull back from her confidence.

'What happened? Mrs Lenister?'

'There came a difficult time,' she said slowly. 'The sisters began to quarrel between themselves. It became very unpleasant – accusations and unkindnesses. It got rather out of hand. And then there was the terrible accident.'

I sat up, barely feeling the pain in my hip. 'You mean Morvern?'

'No, I mean when Miss Gillies had her accident, the burn.'

'Oh.' I sank back against the pillows and my thoughts returned to that very first day when Miss Gillies had positioned herself so that I would see only her unmarred cheek and build from that the template of her good looks.

'And then, Mrs Lenister?'

'Let me tell you all. After Morvern died, the girls

began to turn on each other, fighting in a way that they had never done before. And because Greer's grandmother was considered unsuitable, the Gillieses suggested Greer live here. But Morvern's death had changed Greer and although she hid it, I saw it in her eyes. When the sisters were there and turned their attention away, I could read Greer's anger – and Greer had a lot of anger.

'Things began to happen – a favourite chain of Evangeline's broken, a rip in a new dress, a lost jewel – little things. And always the fault went either to one sister or the other. They accused each other of these mishaps and fell out badly.' Her eyes grew darker, her voice more hushed and nervous. 'But I found her, you see. I found Greer one day. She had a pair of sewing scissors. I saw her cut the dress myself. Greer made the mischief and caused the blame to fall to one or other of the girls. She was responsible for the rift between them.'

I felt a moment of shock that Greer's need to avenge herself had been so slyly and maliciously executed.

'And Miss Gillies' accident? Was she the cause of that too?'

She sighed. 'It was Miss Gillies' birthday and her father had given her a gold pocket watch. It was a beautiful thing and Miss Gillies was very proud of it. She put it away in the library until the party that had been planned. But when she went to take it out again, the face was smashed. Miss Gillies was more upset than I have ever seen her, and she and Evangeline had a mighty row. There was some pushing and Miss Gillies fell on to the hearth in the library. I think she must have knocked herself out, and the hot iron and flames did the damage in seconds. The sound of Evangeline's scream, I will never forget, and Miss Gillies' poor face. That was the end of any friendship between them. They never spoke to each other again.'

'And Mr and Mrs Gillies? What did they do?'

'They too blamed Evangeline, and later we understood that Iskar had passed to Miss Gillies in compensation for her injury.'

'And Greer? Had she smashed the watch?'

'I suspect her, yes.' She turned away and avoided my gaze.

'What? What is it?'

Her demeanour altered. Shadows fell across her face and the air became charged with a sudden fear. She leaned towards me. 'I am going to tell you something that I have never told before to a soul. Do you understand? Not a soul.'

'Tell me,' I whispered. 'I will not repeat it.'

'I found her, you see.' Her eyes were wide with shock. 'I found her that very morning, before Miss

Gillies had even received her present. There was a doll's house in the nursery that Greer loved. Of all their toys, she went again and again to that, but I saw something that day that frightened me. That still frightens me. She did not see me at the playroom door, so absorbed was she in her game. She had taken a little doll and as I watched, she placed its head in the fire. She held it there until the flame caught its hair. I left and did the first-floor fireplaces, and when I had finished I crept upstairs again. She had gone by then but I looked and found the doll in the house, hidden beneath some furniture – its cheek burned black.'

The silence deepened; the breath of something wicked turned in the air.

Mrs Lenister's features were stretched tight in recollection. 'I remember her expression as she did it, caught in the firelight – she was happy.' Mrs Lenister shrank back from the bed and gave me a sharp look. 'I've told you this only because I think you should take care with her and I know that you will not repeat it.'

'She makes you afraid, doesn't she, Mrs Lenister?'

She stood. 'The truth is that she made me afraid when she was only seven and she still does now. I am always cautious with her. Always.' She leaned down. 'Take heed. You are far too careless in your manner when it comes to Greer. I'm telling you, don't get on her wrong side – not if you know what's good for you.'

I closed my eyes and thought of the doll's house and what I had taken – too late, I thought. Too late.

31

Mrs Lenister was barely out of the door when Mary ran in and sat beside me, examining my face with such anxious eyes that in spite of my recent dismay I could not help but smile.

'I'm all right, Mary,' I said. 'Don't worry. It's just bruises.'

I was glad of her company. We played cards until Greer brought up a tray and Mary went to eat with her aunt. I could not look at Greer. There was a hot drink, spiced and alcoholic, and after I had eaten, the pain receded and I was too drowsy to think.

Flames threw swirling shapes to the walls and the house grew quiet but for the creakings and groanings of wood. I fell in and out of sleep, in and out of fear, but my head and body ached almost too much to inhabit it.

Later, I woke to Greer standing over me. She did not speak but glared, her breath warm on my skin. I tried to turn away but she put her hand to my cheek and forced me to look at her.

Her face was stretched with such glittering hatred that my breath caught in my throat. I had not thought it possible to communicate such loathing. I could not move.

She smiled and leaned close. 'You will regret this,' she whispered and held up one of the dolls I had taken. 'You think you can protect yourself by taking the whistle and the figures of you and Mary?'

But in spite of all that I had so recently been told, my fear had reached such a point that no more could be created, and she no longer scared me.

'Get your hands off me.' My voice slurred.

She put her lips nearly to mine. 'Go home. Go back. You're not wanted here. Why do you interfere?'

And in spite of all that I knew of her, and all the warnings, a madness of rage gripped me, shaking my tongue loose. 'You are the one that interferes. You have tried to frighten me from the first. I saw you coming down from the Fiaclach. Hettie learned it from you, didn't she? It was not her that brought her spells and tricks to Skelthsea, it was you. You made the dolls. You summoned Hettie to push me today. You think I don't know that you are like her?'

Her eyes widened a little. 'You know nothing.'

'You put those dolls and pebbles in my room to intimidate me.' My head swam. I could not stop. 'And Mary. You mean her harm. Miss Gillies is too blind to see it, but I see it. Did you kill William? Why?' I hissed.

Her eyes narrowed. 'You don't know anything. And Mary,' she sneered. 'Are you sure she is what you think she is? You're a fool.' She pushed the doll into my face. 'You stupid bitch. You think you know it all. Well, if you know what's good for you, you should leave and take the idiot child with you.'

'You're wicked, Greer. You called her back to do your bidding.'

Her breath rasped and the candle flickered.

'You're a sick creature,' I said.

She lashed out and squeezed my wrist so tight in her fingers that I gasped in pain. 'You know nothing, Elspeth. Nothing. You know nothing and you understand nothing. But know this: I shall make you pay. You will regret this. Forever.'

When I woke again, it was morning. I wondered if I had dreamed the confrontation with Greer but there was a bracelet of redness about my wrist that said otherwise. Although I was stiff and the pain worse, I insisted on getting up.

I sat at my mirror and traced a cut on my neck and wondered what Mrs Lenister would say if she knew about my exchange with Greer. All the time, I fell between anger and dismay. Anger at Greer and dismay that I had fuelled it.

It was church later and I would not be persuaded

to stay behind, so we trailed up through the damp to arrange ourselves on the unforgiving pews. Paterson was there in a new coat, nipped at the waist. When his wife wasn't looking he gave me a lazy smile, but I did not care. Even with the boom of song and the smell of so many living people, I was conscious of only two, not tethered even to Iskar, but who had found us from beyond death. They could come and go when called and there was no door on earth that could keep them out. If they meant harm, they would surely succeed.

Beside me, I became aware of Mary and I turned to her. Her cheekbones were too sharp, her eyes too distant and troubled. It was not me alone who was haunted, and a headache began to pulse at my temple and a wave of nausea rose. The urge to vomit became urgent and I leapt from my knees and fled the church to the curiosity of the congregation. I only just made it to a patch of undergrowth before I emptied my stomach.

I leaned against the wall for support, and when the sickness had passed I realized that for the first time I was alone on the land of Skelthsea – all else were inside the chapel.

It was dusk. A grey and dusty light squatted in the shadows and the sea was an endless black. While the congregation intoned the prayers, a strange silence fell over the valley. My eyes found the ridge where the bitten rock of the Fiaclach lay, seeming older than the island itself, and a hollow boom sounded as an ocean squall found the cave entrances. I had never felt so isolated.

Behind me, from the church came the noise of feet and the rousing swell of song and then the chatter as they spilled out of the door. I stayed where I was, half hidden behind the chapel. Miss Gillies and Mary left first, then the staff. Mary looked around for me and I was struck once again by the inflexibility of her looks – as if her features had been stamped upon the skin, not giving allowance for a smile or a frown.

Miss Gillies chatted to the villagers and Mrs Lenister and Greer talked, although Mrs Lenister plucked constantly at her coat, adjusting its weight. Her cheeks were red with cold. When I looked again, Greer stood by herself, her headscarf tugged by the wind. As I watched, a lantern swung, capturing her expression, and there was something that stilled me, something new in her eyes that I had not seen before. She was a woman transfigured. For those moments, she was almost beautiful. I was mesmerized and followed her gaze.

Paterson. He was laughing now. It was deep and warm. A couple of village girls blushed and smiled, but every now and then he looked sidelong and fixed on Greer. Did he see it as clearly as I? I recalled Paterson's words of a few weeks ago when he had suggested other alliances at Iskar – Greer? And Greer's look now spoke not of some light affection but something that burned hot inside her.

How long had she felt that way? Had it been before Hettie began her affair with him? If so, what must she have thought when her friend – a friend who could so easily have turned the eye of any man on the island – began her relationship with him? And had Hettie known of Greer's infatuation before embarking on the liaison? If that had occurred, would Greer have still loved Hettie at the end? How must she have felt if the man she loved was under the spell of her beautiful friend? I let out my breath. Perhaps I was allowing my imagination to roam too liberally. The only thing in the matter that I knew for certain was that Greer had given her heart to Paterson.

It was a relief to take Mary to her bed later. At her dressing table, I caught a glimpse of myself – bruises blooming and behind that the force of my fear – and I barely recognized myself. I could not bear the thought of the coming night and even then, although I tried to deny it, somewhere in a distant room, a faint whistle seemed to shiver in the air. Mary was restless and seemed reluctant to settle. Twice, she put her hand on my arm and opened her mouth. Finally, as I tidied away, she pressed a piece of paper into my hands and I felt a sigh of impatience. I did not have room to consider her needs; my own were too pressing.

'Please don't leave me, Elspeth,' she had written. I bent and kissed her and saw how my refusal to engage made her withdraw. But I did not have the resources left to attend to anything other than my own anxiety and all I could do was reassure her, once again, that I would not abandon her; that I had no intention of leaving Skelthsea. I turned at the door to wish her a final goodnight and she regarded me sadly, not even mustering a smile.

I could not settle and worked myself up into a frenzy of apprehension. My thoughts became muddled; I thought I heard the murmured voice of the doctor outside my room in Circus Gardens, the tapping of the ivy. The wild, unfettered rush and retreat of emotion battered me.

In the candlelight the room shivered and I thought with sickness of Hettie's lullaby and how it had become a thing of such dread.

Eventually, I returned downstairs and found Mrs Lenister cleaning the kitchen and pleaded a headache, and a little later she came with a bitter drink. I must have fallen asleep because when I came awake, the room was in darkness and I was still fully clothed.

A draught razored in from the door which now stood open. My head was thick. But under the sleepy weight of what I had drunk came the awareness that something was wrong, very wrong. I wanted to retreat back to the world of dreams but consciousness ripped me full awake. There was a pressure on me and a scent that stifled my nostrils. Whatever Mrs Lenister had given me slowed my reaction and it was a second or two before my brain unravelled my senses, and when it did, my heart leapt out of my chest: she was there, across my body, a weight of dead skin and bone that lay along my own.

32

Terror robbed me of all reason, robbed me of movement, and I could barely struggle against the burden of her weight. The room was too dark to see, but the smell of her was suffocating: sea and rot and rage.

I opened my mouth to scream but the scream that was in my head came out only as a choke.

Standing, I fled breathless into the corridor, the imprint of her body burning my skin. My heart could not manage such terror. It was only as I regained reason that I saw that Mary's door was ajar and discovered her bed was empty, her sheets cold. Fear struck deeper than a blade.

The house was stamped with silence; all sound seemed sucked from the air, leaving a twist of something abhorrent. I ached to hear Mary, to hear her in the trap of a nightmare somewhere beyond, but her absence rung in the shadows. She was not here.

My heart hammered harder and I tried William's wing, but it was locked. In my own room I lit a candle. Upon the bed lay Bobbity – the new dress half torn from her body and soiled in moisture and mud.

All I could hear was Greer's warning. I fled to the main hall where a drift of dead leaves lay across the tiles.

Oh God. What had my actions caused? Greer had said that I would regret what I had done forever. Hettie had taken her. Too late, too late. Grabbing lantern and boots, I rushed into the night.

It was icy. Sleet burned my cheeks. Below me, the sea glittered like a half-closed eye and the vision I had had that day in the valley returned to me: Mary falling through the darkness, suspended by gravity, hair twisting in the pull of the tide. Was she destined to be found as Hettie had been, washed up later, if at all?

When I was far enough from the house, I screamed her name but the wind grabbed my words. On the sands I scanned the water's edge, called her again and again as I stumbled along the strand. Then something made me lift my head to Stack Mor.

There in the circle was a flicker, and some instinct gave me certainty. Surely this was where Hettie had taken her to be thrown from the rocks to die as William had, and the belief gave direction to my feet. The valley was in shadow but I began to race up the path, my face and hair wet with rain, dress dragging. Twice, I slipped, and my lungs hurt with the pain of finding breath.

Near the graveyard I began to call again and glimpsed a figure on the Fiaclach but cloud passed over the moon and it was lost. What if I was too late? I screamed her name until my throat ached and when I reached the circle, it was empty.

I made for the rocks and the lantern caught a splinter of colour. It was the ladybird pin for Bobbity, bent out of shape. My search was desperate then. I was at the point of considering all lost when finally, I saw her white foot lying still amongst the grasses. The pale arch, slick with moisture. I was too late.

She lay on her front, gown and hair black with water. I pulled her into my arms, lifted her face to mine. Her skin was as pale as fish belly, her eyes closed. I put a hand to her chest and my cheek to her lips. Please, I begged. Not again. Not again.

The last time I had prayed it had been for Clara, but He had not saved her.

Now, I prayed once more. I prayed for forgiveness. I prayed for help and I prayed for mercy and most of all I begged Him that Mary would live.

I was aware of nothing but her face against mine and that unbearable yearning. Then, barely noticeable, came the soft feather of a breath. I shook her and rubbed her back, willing her to come awake. Finally, her lids opened in dazed shock. Relief thawed my senses to the burning cold. And with my arm half supporting her, we somehow reached the beach.

By the time we entered the house, every muscle

ached and I was shivering violently. I got Mary to her room where I stripped her of wet clothes and wrapped her in blankets. Meanwhile, I refilled the bed warmer with the remaining embers and when that was done put her beneath the covers. I changed quickly and in the kitchen made a hot drink with whisky and sugar and brought up a bucket of peat. She was so still that I thought that she was dead, but she came half awake when I pulled her up to sip.

She fell between waking and sleeping but the drink brought colour to her pallor. I re-made the fire until it leapt in the grate. I warmed my hands and rubbed her feet, and soon she began to shiver and the dreadful sleepiness receded. I got into bed beside her and wrapped my arms about her waist as though it might keep her from wandering. Her skin was warm against mine. I knew in my bones that she had nearly died. So nearly. And I also knew, in that moment, that it was not just a sense of loyalty that kept me at Iskar, nor the memory of Clara's courage – but that somehow, woven into the days of loathing and fear, something had crept up on me, as quietly as winter to spring itself until, without warning, buds are springing from the empty branches and snowdrops pushing through the slumberous earth.

I loved her. I loved her like my own. And I put my cheek to hers and wept.

33

Mary woke before dawn, scrambled in a nightmare. Her head was hot to the touch and her chest wheezed. I went to my room and removed Bobbity, cleaning as much of the mess from her as I could, then I hid our wet things. It would not help Mary's cause if it was believed that she had been sleepwalking again. When dawn broke, I went downstairs and found Mrs Lenister.

'What has happened?'

'Mary has a fever.'

'And you? You look terrible pale. Get yourself to bed, Miss Swansome. I'll be up in a moment.'

Mary lay moaning in sleep and when Mrs Lenister came with a cup, she put a palm to her forehead and shook her head. Gently, we woke her and gave her the tonic. Mrs Lenister watched as I drank mine and I lay back down beside Mary.

'I shall sleep here and watch her. Would you get Greer to make me up a cot?' My lips recoiled from the name.

Mrs Lenister nodded and hurried from the room,

returning later with a clean cloth and a jug of water steeped in herbs. I watched as she laid strips of willow bark across Mary's head and bade me do the same.

Greer came soon and I steeled myself to meet her. I wanted to gauge her reaction but her face gave nothing away. When she reached the door to go, Mary moaned, throwing the covers off, and Greer paused, but her expression was unreadable.

I slept then, long and deeply. When I woke it was to the blazing fire. A bowl of oil and herbs stung the air.

I do not know how much time passed until I came round again. Although I ached, I no longer felt feverish, and with relief, I found Mary sitting up, a bowl of bone stock on a tray before her.

Mrs Lenister sat by the fire and a smile split her face.

'What time is it?' I asked. My throat was sore.

'Late. You slept for twelve hours.'

'Will Mary be all right now? It looks like the worst has passed.'

'She will. You both will.'

The following days were dreamlike. Although the fever receded, it left some mark of distance on me, but beneath that a new fear had begun to drum – because I knew for certain that Hettie not only made changes in the physical world but that she intended the worst kind of harm. I stuck to Mary like a shadow, not leaving her except for the times she was with Miss Gillies, and even then I stayed close in the drawing room, my attention half turned to the corridor in anticipation of her step. Staying had become not just terrifying, but dangerous.

I had to leave and I had to take Mary with me. Yet I could not imagine how that would be achieved, and soon my head began to hurt again, and in spite of the fire, I could not get warm. After that, the hours began to blur. Sometimes Mrs Lenister's face was there, at others I saw Greer bending over me, and once, Miss Gillies standing at the window of my own room although I did not remember moving back.

At times I woke from nightmares with Hettie laid across me; others were full of Clara or Mary who had become lost and I could not find them, and when I woke it was to the pain of knowing that at least one would never return. Eventually, my head cleared. I found Mrs Lenister asleep in the chair beside me. She stirred when I sat up.

'Is Mary still well?'

She looked at me with relief and touched my brow. 'Much better,' she said. 'It is you who has been ill.'

'I've been ill?'

'Yes, you took a turn. Been worrying us all half to death.'

'How long?' I asked.

'Three days.' She came and straightened the bedding. Her movements were nervous, and when she had finished she did not go but pulled at her sleeve and watched me.

'Is everything all right?' I asked. 'What is the matter? Mary is definitely safe?'

'I told you she was. We've all been managing but we can hardly keep her away from your room. She watches you like a personal nurse. You have captured her affection, Miss Swansome. It gladdens my heart.' She smiled. 'Let me get you some broth.' She went to the door but at the handle she turned. 'It's good to see you back, Elspeth.'

Elspeth. She had not used my name before and somehow its sound on her tongue stung my eyes.

I lay back against the pillows and listened to the sea and the wind. It did not seem many moments until she was back with a bowl and a drink on a tray. She helped me up and placed it on my lap.

'Go slowly, your stomach is not used to food.'

Her eyes watched the spoon as it went from bowl to lips, and with each mouthful, I read her approval. Afterwards she took the tray and laid it on the table and then fussed at the pillows and covers, but in all her movements, there was some urgency of purpose beyond the taking care of me.

Finally, she turned from straightening the curtains. 'You asked me once if there were ghosts.'

Her weight creaked as she sat beside me. 'I hardly dare say what I am going to say, but I feel I owe it to you. I have seen your fear and I have not taken a share in it. You deserve better from me. So I'm going to tell you what I have felt.'

I was not alone and I could have wept with the relief of it.

'I smelt her perfume.' She twisted her hands nervously inside each other. 'Hettie's scent – it was always the same, always. After she had left, I would come across it in the air. But I thought it must be my imagination – or perhaps – I don't know what I thought then.'

I recalled the stale odour that sometimes accompanied Hettie's visitations. 'How often did this happen? Where were you when you felt her?'

Mrs Lenister looked wary. 'From time to time. I may have been taking feed for the hens, cleaning the silver or cooking, and it would be around me and a strange feeling.' She put a hand to her chest. 'Then there was that time you asked me about your shoes.'

I recalled that first morning that I had found them that way. The curiousness of it.

'That was her way. She always did that. We used to

tease her about it. When you told me – I knew then that she came to you also.'

I felt dizzy.

'Mary. Mary sees her, doesn't she?' Her cheeks were flushed. 'Hettie. And William, too. I've seen you watching her – chattering away to the air. Miss Gillies, how she hates it. At first, we all thought it was an imaginary friend to make up for William's going, but she's too old for them, isn't she?' Mrs Lenister leaned so close that I could number each blunt eyelash. I felt her fear.

She sat back and gathered herself. 'When I first began to smell Hettie or hear her step – her quiet, quick way of walking – I told myself it was my imagination because I did not know then. Did not think –' Her face creased with fear. 'We had not learned that she was dead then. We believed that she was on the boat or making trouble elsewhere.'

For a moment I thought that she might cry. A shadow passed over her features.

'And then, that terrible morning – on the beach – and we discovered that she had been dead all that time – all that time.' Her skin was ashen. 'And even after that, when we knew she had perished, I could not bring myself to acknowledge that she had returned.'

I shivered.

'What does she want?' I whispered.

Mrs Lenister frowned. 'Want?'

I turned away. The thin light brushed the walls. 'When I was on the ridge. I was pushed. I heard her. I felt her and I was pushed by her.'

Mrs Lenister flushed, rested a hand on my arm. 'Don't let it carry you away, Miss Swansome. What are ghosts, after all? Just air.' But there was something in her eyes that made me think she did not entirely believe her own statement.

'Since my arrival,' I said, 'someone has come by my room nearly every night and hummed a lullaby. At the beginning, I believed it was Greer as she turned down the lamps. I would hear her approach from the main hall and when she reached my room she stopped and waited there.'

'It was Hettie?'

'It was not someone living.'

Her face fell.

And so quietly, like a whisper, I hummed it, and in my voice for an instant I heard the dusty timbre of hers.

Mrs Lenister paled at its melody.

'You recognize it, don't you?'

She nodded slowly, then turned and went to the window. 'I heard it only once. Not long before Hettie died, a baby fell ill. I went to help. Just a wee thing it was, not a day old. Hettie was already there when I arrived. She had the child in her arms. She was singing that tune.'

'What is it?' I said. 'What is the song?'

She shrugged. 'Some tune from England. I had never heard it before. She was a wicked, complicated girl but there was something about her that day. Gentle, I thought.' Her eyes were sad. 'I thought – I don't know – she held the baby, like it was her own.'

She collected the tray. At the door she paused. 'I do not believe that Hettie has the power to harm you. I hope you don't leave, Miss Swansome. You are good for Mary, but you should take care. The island knows that you have questions – questions nobody wants to hear. Remember what I said about Greer. If you really were pushed you should look to the living and not the dead.' And with a bob, she moved out of the door.

34

Gradually, I returned to full health, but it was slow and the knowledge of the danger to Mary weighed heavier even than Hettie's dead step or the sound of the whistle. Hettie came when she chose; I could not stop her. Most nights, I would lie beside Mary in the grip of horror until, after a pause outside Mary's door, her steps and melody drifted away.

I turned the possibilities in my mind. Could we find a way to take the boat when it next came? Could I make friends with one of the fishermen, beg him passage to the mainland? But the idea that someone would keep such a secret, would undertake such a venture, seemed impossible.

I began to visit Bridget more often again. In spite of her open manner and gentle questions, I found I could not speak of the things that preyed on me. I had heard the way she talked of ghosts and witchcraft and she would think less of me for believing, and how would that make her feel about me as nanny to Mary – Mary who was already so blemished in the eyes of the island? She could see I was unhappy and she showed her friendship in the gentleness with which she treated me, and afterwards a little dread was always temporarily eased.

It was one such afternoon, when I was tidying myself after a visit to Bridget, that rifling through the drawer, my finger caught on something sharp, drawing a bead of blood. I investigated and came across the dolphin brooch nestled among my clothes. I clearly recalled putting it back into the lining of the curtain. I opened the window but could not, at the last moment, bring myself to dispose of an object so lovely.

Once again, I found myself in Hettie's room. From the window the island and sea merged in a marriage of blue and grey shadow. I felt at the curtain but the letter remained. Who had come and taken the brooch? Hettie herself? And I wiped my fingers down my dress.

In my room I laid out the pebble and dolls. I pondered again the strange warmth they gave. Had they been manufactured before the deaths to bring death itself? Or to summon them from death? Were Mary and I to die too? And then the thought came to me. It had not occurred to me before: if the dead could be drawn from their graves, could they also be returned there? And I felt a prickle of hope.

Before lessons came again the next day, I collected all the objects I now associated with spells and put them in a bag. It seemed to take an age before we were draining our coffee cups and Miss Gillies was checking the drawing room clock.

'Lessons already,' she said, as she did every day.

She barely glanced at me as I took my leave. In spite of the sun, there was a grating cold to the air and even the gulls that perched on the flat rocks huddled in their cloaks of feather.

My boots crunched on the shingle as I made my way up the valley through the trails of peat smoke and the mewling of gulls. Ailsa was in the kitchen and once again I had the uncanny sense that she had expected me. Her face was drawn and for the first time, as she pulled her shawl closer, I recognized that she was truly old.

'Come in.' A wisp of smoke escaped her lips and she closed the door behind us, beckoning me to the bench in the kitchen. The usual array of items littered the table and a fishing net that she had been mending fell in folds to the floor.

'So?'

I was not welcome.

I took the pouch from my skirts and emptied the contents on to the table. Her fingers tightened on the pipe.

'You bring a widows' whistle here?' She was aghast.

I thought of all the scorn that must have leaked from me when she had first shared her information and a flush of shame rose on my cheeks. 'I didn't want to believe it then,' I said. 'Even though I suspected then that Hettie and William had returned. I wanted to think it could not be true. Now I know otherwise.'

Her eyes flashed up and she took the pipe from her mouth. For a moment I thought she was smirking but it was a grimace. The widows' whistle was repugnant – thin and discoloured with a roughly hewn mouth. I tried not to imagine the sour, rotting air that passed through its tube.

'I wonder who called them. Just as you must have wondered,' she said finally and reached out for it, withdrawing her hand at the last instant. 'Was it Mary?' She spoke softly, as if to herself. 'Perhaps the desire to see those she loved again brought them both, or perhaps it was for another reason.'

'I don't believe they were called by love. They were returned to continue their evil.'

Her eyes went wide at my vehemence but she made no comment. The first spatter of icy rain hit the glass. She picked up a piece of wood and pushed at the dolls and pebbles. 'And all these were at Iskar?'

I nodded. 'This is not Mary's doing. I believe Greer has a hand in it.'

'Greer?' She paused. 'Perhaps.' She raked my face and for the first time there was pity there.

'Can you help me? If there are spells and tools for summoning, there must be those for dismissing. You seem to understand these things. We must send them back to their graves. She means harm, harm to Mary.'

And I imagined Iskar without its ghosts and with sleep uninterrupted by Hettie's winding voice or her presence, and was flooded with longing. Without the ghosts, Mary and I would surely be safe and Iskar would be tolerable. I did not believe that Greer would have the courage to harm Mary by her own hand.

She frowned. 'And Hettie means you ill too?'

I laughed. The very smell of her, the sinuous undulation of her lullaby . . . I remembered her at the graveyard, the feel of her weight before I fell. 'Yes,' I said, 'she comes to terrorize me, to taunt me and to harm me.' My hands were gripped tight on the table and I released them self-consciously.

I smoothed my skirts and began to pack the pebbles and dolls away.

She put a hand on mine. 'Wait.'

Getting up, she rummaged in her coat which hung from a nail by the door and drew on some gloves, then she returned and examined each doll, each pebble.

'These,' I pushed at the images of myself and Mary, 'I found these in the doll's house to which, I think, they originally belonged.'

'Ahh.' She nodded to herself. 'These are best destroyed now before they can do harm.'

'They have not already done so?'

'You are both alive and well? Leave them here. I shall make sure they are disposed of without injury to you or Mary.'

She hesitated before picking up the widows' whistle with a tremor. When she looked at me there was anger. 'These are dangerous tools.'

I did not like the way she said it. 'Can you help me?'

'I'm not sure. I understand a little of these crafts. I know how it would be done, but I do not know if it would work.'

A flutter of unexpected hope rose in my chest.

'And you would do this – for me?'

Her eyes were unreadable. 'For all, Miss Swansome. To have the dead return to the living upsets the laws of nature.'

I nodded.

'If I am to help you, you must also be there.'

'Anything,' I said, 'I will do anything that is necessary.'

'All right.' She turned to regard the failing day; frost was beginning to crystallize in the air and she laid her hands on the table and gazed at them. I sensed, for the first time, some uncertainty. 'It's the full moon tomorrow. It would have to be then.'

'What do I have to do?'

'Meet me here at quarter past eleven. Can you get away unseen?'

I said that I could.

'Bring anything else you think might be part of it and then we must go to the Fiaclach.'

Mary was quiet during the evening. I wanted to tell her what Ailsa had planned. I wanted to rejoice with her that soon we could send William and Hettie back to their graves, but she was restless. At one point I found her outside William's wing, standing quite still, her hand upon the door. When she saw me, she started and guilt scurried in her eyes. As we walked back along the corridor, she turned to gaze up at the attics and I felt a wrinkle of disappointment and concern. I wanted to believe that she felt as I did, that she wanted them gone, but I was not certain.

I put Mary to bed and recalled the icy clamp of her skin as she lay white and wet amongst the frozen grasses, and I longed to hold her tight, hold her forever. As if sensing the direction of my thoughts, she reached out and touched my cheek and there was an expression of such sadness on her face, a look that did not belong to a child, almost as if she were the mother and I was the infant in peril.

'What is it, Mary?'

She got up and went to her paper.

'The dolphin,' she wrote.

My heart quickened. 'What do you mean, Mary?' I thought of the brooch. 'Was it you?' I asked. 'Did you move it?'

She frowned and wrote again, 'Don't go near it. You must stay away.'

'What about the dolphin? Explain, Mary.'

But she laid the pen down with finality.

I stood, uncertain, then leaned down and kissed her cheek.

Mary's comment was odd. Why would it be necessary to stay away from such an object, and how had Mary come to have that opinion?

But I did not have time to ponder. Because, on the morrow, Ailsa would rid us of Hettie and William's ghosts forever.

I stood at my window. Ferns of frost curled against the glass, and behind, the black vault of sky held a moon nearly at full circle. I almost willed Hettie's soft step outside so that I could rejoice that soon it would be over.

35

The next day I spent in a fluster of nervous excitement. Later, Ailsa and I would venture out to the Fiaclach. For the first time I felt that I held a weapon against my fear.

All day I was like a wild cat held only with a satin thread. When the time came and I found myself in church, I knelt on the floor and was assailed with a hammer blow of doubt. The wooden cross hung above the altar. How much everything had changed, and my beliefs altered. I wondered how my father might have reacted to what I planned to do that night. But those days were so far behind and the memories so in contrast that I could not hold on to the thought.

Outside the church, I shivered. The moon seemed to pin the sky to its backdrop of stars and it was still. So still. Even through the rasp of the sea and the chattering of voices, it was as if the Earth had ceased its turning.

Mary slipped her hand into mine. Her eyes in the moonlight were inscrutable. Although she did not speak, I was aware of the avalanche of words that were dammed up behind her lips; they pressed against me and I gained the sense that she knew my thoughts or somehow what I planned.

That evening, all sounds seemed amplified, or perhaps it was the heavy silence behind them that gave that illusion, but during supper I was overly aware of the chink of silver on china and the noise of a glass as it was replaced upon the table, the rustle of skirts and napkins and the clock like a gunshot.

The evening dragged, but finally I was able to take my leave and go to my room where I dressed warmly and waited. From the window, the moon had grown larger. Soon, the sounds of the house abated to the humming eaves and creaks of old boards. In the corridor damp hung about the lamps like a white breath. I stopped and listened – nothing but the night's hush; I listened deeper for the sound of Hettie but met only silence.

I did not need a lantern and I laced my boots and left by the chapel. Outside, the stillness had grown as if, like me, it too was waiting for what was to happen. Ailsa was at her door, her eyes bright in her face. She carried a sack and handed me a bag, knotted at the opening.

'Take care with it,' she said. And as I took the bag, I knew, somehow, that it contained the dolls.

We did not tarry. She led me past her cottage and along a path that banked the north side of the island. Waves rose and roared below us.

The moon was falling behind the ridge but eventually we climbed again to where Stack Mor reared into the sky. Ailsa's assured step led us through the narrow channel between the rocks and soon we were out of the wind and into the rush of sudden silence.

The Fiaclach lay like a scattering of old bones. And I was afraid – afraid of the strange magic of Skelthsea, of the way the Fiaclach repelled me and of the rattle within the bag at my back, as if the contents had come alive and the dolls had opened their eyes to watch me through the cloth.

In the cold whip of air was the smell of the earth – millennia of leaf mould and seabed – and I imagined the scuttle of prehistoric life along the floor of the ocean in the winking blackness, the strange shapes of never-seen creatures, and felt the energy of something beyond my knowledge.

We reached the stones. 'Here,' she said and crouched by the altar.

She emptied her sack and laid the contents on the ground: candles, a small bottle, a knife and a piece of black cloth.

I watched, half appalled, half fascinated whilst the moon sat unblinking in the cloudless sky. A tremor went through my fingers as the cord of her bag nearly fell from my hand and I gasped. Ailsa looked up, a flash of alarm in her eyes.

Finally, holding her scarf about the candles, she lit them and placed glass chimneys over their wicks, but even then, they flickered wildly like creatures seeking escape.

From the bottle she tipped a little liquid on to the slate where it shone and a strange, unpleasant scent made its way into the air.

'Elspeth . . .' she had been talking but I had not heard, 'it's time. Pass me the bag.'

It jerked as I picked it up, as if whatever was inside had moved. For a moment she did not open it but let it rest on the ground. Then she closed her eyes and began muttering. Gently, she took off her gloves and eased the knotting and, reaching in, she drew out a pebble and placed it at the corner of the stone.

'You must turn away, Elspeth.'

I obeyed. Behind me the clatter from the sack and the sound of a click – a rustle. I tried not to listen and concentrated on the stark shapes of the cliffs ahead.

'Turn back now.'

I turned. The altar stone was covered with the cloth and beneath that were the humped shapes of the dolls. Ailsa handed me the widows' whistle. With horror, I wondered if she might ask me to blow it.

She began to whisper fast and furiously, and I was reminded of the first time that I had come across Mary in her nightmares and the senseless gibberish that fell from her mouth.

As she spoke, I became aware of some other voice that, although it was outside me, I seemed to hear with some inner ear, because when I listened it was only Ailsa's that was clear on the night. At first that voice was unintelligible, but gradually it grew more persistent and I felt its will, resisting Ailsa. I grew sick and dizzy, putting my hands to my ears to block the sound, but still it continued and began to squeal and scream like a rabbit in a trap. The candles, which had grown still, began to writhe and the night opened like a mouth.

Sweat pooled on my back.

'Pass it to me. Now.' I watched, horrified. It looked, at that moment, as if the widows' whistle was no longer bone but a piece of flayed skin that shrivelled in the moonbeam. Light played on its surface until it seemed to wiggle. For a second, she paused and her eyes met mine in abhorrence. The air gathered around us and in its mystery, I knew that we were watched from some other world. Her hand resisted and then she reached out and took it in her fingers. I could not breathe. Slowly, she brought it to her mouth and the world spun.

Repelled, her lips closed over the reed. She did not blow but drew in her breath, pulling a sound from the widows' whistle that I had to strain to hear, a sound that came from somewhere beyond the realm of mortal reach. It ripped, for a moment, that membrane between life and death. I knew it in my very soul. Something unearthly screeched from beyond the rocks, as if it had taken the opportunity to slip from another kingdom and pause a while. Then silence. Beneath the cloth, the dolls jerked. And then lay still. The candles fell to quietness. For a flicker, our eyes met, then she lifted the widows' whistle high in the air and snapped it, leaving the sound to bat the air.

Ailsa leaned over, spat and wiped her lips with the back of her hand. It was over. One by one she quenched the candles. Last was the cloth. Gently, she lifted it and there they lay, still as the night, two wooden dolls in clumsy dress. They no longer held my horror; they were but pieces of wood. They could not walk, they could not speak. They held no souls.

From the bag she took two small boxes, lifted the lids and placed one doll in each, then threw in a handful of earth. With nails and a hammer, she closed them shut. They were coffins, two doll-sized coffins to hold the dead. Finally, she stood.

I followed, aware that the wind had grown and that cloud was skimming the sky as if it scabbed the wound that had ripped it open.

At the north beach, the water had fallen quieter. We crossed a sheet of granite where waves banged at the lip. Ailsa took out a handful of sand and poured it into the bag, secured the top and flung it to the ocean. It made no sound. They had been returned and Iskar was empty. Empty of the dead.

I did not need to ask; I knew that the ghosts had gone. It was over.

PART THREE

Sleep, he said, lay down your head
No monsters here beneath the bed
And so she slept star-deep in dreams
Moonlight on the window panes.
And did not see it standing by
Or how it watched with its dead eye.

Anon

36

I passed down the valley as if in a dream. The chimneys of Iskar were smokeless, the windows empty but for reflection, and I thought of my own room and the fire that would surely have fallen cold and the corridor with its precarious light that would now remain unwalked but for the living.

I entered through the chapel door, aware of the way my boots clicked on the tiles and the damp smell of the air. I passed by the library and into the hall where the aroma of supper was turning stale.

I paused. It was so still. They had truly gone, to lie in the cold earth, while time ate away at flesh and bone, and my heart sagged with relief. Removing my gloves, I placed my hand on the banister and began to climb. At the top the corridor stretched without malice and even the patch of darkness that fell from the garret stairs did not make me afraid. Wind whispered cleanly through the eaves above, empty and untainted but for rotting wood and the crumbling nests of summer birds.

I paused to look in on Mary; her face so peaceful

that I imagined she too understood the ghosts had been returned to death. In my own room, the curtains were open and the moon was beginning to sink behind the ridge. Its bright skin had dulled to a pale yellow. Below, the sea lay nearly motionless and somewhere on its bed, the bag of dolls and pebbles rocked against the tide. Dead, I thought. They are dead, and although it was early, I was not tired but wide awake.

For some time I lay between the chilled sheets, my thoughts revisiting the night, and flashes returned to me, the twitch of the dolls beneath the cloth and the strange, other-worldly sense I had experienced, as if I had been reborn into a different landscape.

I woke to the sense that the house felt different and I hummed to myself as I washed and dressed, pinning my hair up with extra clips. I thought of Clara and I was proud. This time, I had thrown myself into the heart of the dog fight. And Mary, this must work in her favour. I imagined the day now that she would speak, no longer haunted by ghosts. And that soul-piercing sight I had glimpsed in my head of Mary in the water and the words scrawled on a piece of paper *I am going to die*. Without Hettie to do harm, that too must surely hold no more threat.

Mary lay asleep, her face as smooth as cream, and a shiver of joy went through me. What could Hettie or Greer do now? She stirred and sat up, her hair as cloudy as sea foam. With all the horror banished, I thought, what other reason might there be for Mary to remain mute?

'Good morning,' I said and her eyes opened wider to my happy tone. As I dressed her, I did not sense either loss or gain. Did she realize? Did she feel their absence? On the way downstairs, I was struck with elation and I wanted to laugh. Even Greer, as she passed with her spiked looks, did not touch me.

The days grew brighter and although I was aware that Mary occasionally paused and searched the space around her, as if looking for something that was no longer there, I did not pay it much heed. She would learn soon enough. More than once, at the beach, her gaze would slide towards Gulls Cry and a crease of anxiety form on her brow, but I kept my counsel and soon she stopped, as if she finally knew that they had been banished forever.

We spent hours on the sand and began a shell collection for which Miss Gillies promised Angus would make a frame. Returning from the sea, we sat at the library table until we were stiff, fixing shells to backdrops of black velvet and poring over the library books to copy down the Latin names.

Sometimes, I would glance up and realize how

much time had passed and that shadows were gathering. We would smile and I would stretch my aching back, and we would find Mrs Lenister and demand tea and bread.

Later, I would read to her or make up stories, and all the time I worked at drawing out her voice. She would speak in time, I told myself.

The days became the sort I had imagined so many months before in Edinburgh. Mary and I took to the ridge, running between the icy grasses, pausing to catch our breath, her cheeks pink, her eyes naked to pleasure. When my heart paused on my sister, it was with an ache that was somehow easier to bear with Mary beside me; it did not always come with that deep sting that took my breath away. In Mary, I had found more than love; there was redemption. When I caught her smile or pulled a laugh from her, saw how she hung on every word I uttered, I could have cried for the love of it. And it filled that tender space where Clara had been, with a new hope.

It had been a week of heavy rain and we had not been able to spend much time outdoors, but the morning rose to a clear sky. I could hardly wait to be in the fresh air. Mary took my hand and tugged me to the path that led to the glittering north beaches where we caught fish and then huddled against the rocks.

She laid her head against my arm. The island felt free, just wind and air and gull sound, natural and wonderful. When we climbed back, hungry for lunch, my heart was happy.

Lunch brought Robert and Bridget Argylle for a visit. Mary and I went up to change and tidy ourselves, returning to the anticipation of warm company.

'You are very fortunate in having the wonderful Mrs Lenister,' Robert Argylle said, helping himself to fish.

'Mrs Lenister is an asset.' Miss Gillies looked pleased. She wore her best jewels and the flowered dress I had seen when the doctor visited. Fires kept out most of the cold and the dull shine of beeswax on the furniture gave it a new life; it was as if Iskar itself had been released from some curse.

'It is good to see you looking so well,' Bridget said to her friend. 'Things are finally settling down after these terrible earlier months.'

She turned to me and whispered, 'And you too. I have thought on your recent visits that you had seemed troubled. Today, that is not so and Mary looks well too. Credit must be due your way.'

'And how do you find us now?' Argylle asked me. 'Now that you have been here a while?'

'The island is very beautiful.'

'Wait until the spring and summer and you shall see it in all its glory then.' But he looked away quickly and I knew that they all expected me to be gone by then.

'How do you communicate with Mary?' Bridget leaned in close to ask me this.

'We write notes,' I said, 'don't we, Mary?'

'Such an obvious thing,' Bridget said, 'although I do not know that I would have thought of it.' She gave Mary a warm smile. 'And what do you tell in these notes of yours?'

I thought of the secret conversations that we had had and was grateful that Mary could not say.

'We talk of everyday things, although words are not needed as much as you might think. We have an understanding.' I gestured at Mary.

They all watched her curiously and this time she nodded and smiled. A look of astonishment came over Miss Gillies' face. My heart soared. Against the conquering of the undead, the conquering of Miss Gillies' prejudice was nothing. If I could fight ghosts, I could fight to keep Mary at Iskar.

Too soon, it was time for the guests to go and Bridget pressed her palm to mine and leaned in. 'Promise that you will visit me very soon. I have missed you.' And then in a whisper, 'Well done with Mary; I have not seen her grin like that in a long while. I had not thought it possible.'

The atmosphere left its goodwill upon the air and

Miss Gillies was lively and chatty, engaging even Mary in her smiles.

It grew colder. Winter had come fully and laid its hand everywhere – on the trees, stunted by wind, on the rocks and grasses; even the sea seemed cast in ice. Greer left me alone now – her ghosts had gone, and with them, some of her rage and power. Hettie was not there to instruct, and if she suspected me of banishing her spirits, she did not show it and I rejoiced.

I should have been happy. For a time, I was. The nights brought no fear. No pebbles or dolls were left in my room, but something began to prick at me. And one night as I lay on the sheets with the sound of the wind and the crack of cold in the joints of the house, a new sense crept upon me – a sense that I had missed something. It was as if I had dropped a stitch and now the fault was knitted permanently into the fabric. The doubt lingered, and while the days still passed without event, the nights became defined by uncertainty. Sometimes I would lie alert; at others I would fall into uneasy dreams and wake, skin icy with fear and a voice in my inner ear – *wrong, wrong, wrong.*

I told myself that it could not be so, that there was nothing left to conquer other than Mary's own silence, but at the back of my mind there wove something nervous like an anxious finger tap. I tried to examine that tiny bell of alarm but I could not fathom it.

Patches of dried skin appeared on my arms and wrists and tormented me with itches; the bodices of my dresses grew loose.

One night I fell into a vivid dream. A dream of oceans, kittiwakes and cliffs that caught the sun, but the scene soon changed and I dreamed of Hettie once more – a dream with the bloody rip in her belly. This time, I was walking through Gulls Cry behind her, through the sand-strewn hall and up the stairs to the first landing. All the doors stood open, lit by candle-light. Up we climbed, up into the gathering darkness, while below, the candles went out one by one. Hettie paused outside her bedchamber, then made her way to an adjacent room. She stopped and turned, her face a blaze of fury, then she reached out one slim arm that seemed to slither through my skin and reach inside my chest, sending me into sick, shivering wakefulness.

I sat up with a gasp and climbed from the bed. The house was still. She was not there. I gazed out of the window to where the sickle of white moon turned the rim of the sea to silver. Moonlight picked out Gulls Cry too and I remembered the dream; its clarity and how the candles had snuffed to blackness, one by one.

The next morning the dream nudged against me

and some doubt became dislodged. As we passed Gulls Cry, I felt a strange pull. I had no desire to return. But even as I told myself this, it was as if someone was beside me, shaking their head in silent warning, and somehow I knew – I knew that I must do so.

I could think of little else. Afternoon came, spilling shadow into the rooms and bringing a biting chill from the window. I hoped to make an escape at lessons, but Miss Gillies was called away on island business. Time dragged, the hands of the clock seemed weighted with lead. As I climbed the valley for the service, I felt half-ghost myself.

In church, I knelt beside Mary and then we were returning, and I began to regret what I planned. When we entered the hall I was shivering in anticipation.

Dinner passed. I put Mary to bed and then sat with Miss Gillies in idle conversation until I too climbed the stairs. In my room I waited, curtains open to a moon smudged by the glass's condensation. I waited one hour by the clock until, finally, I returned to the corridor and to that unsteady light cast by the lamps. It was silent, the energy turned down, and tiptoeing to the balustrade, I leaned over to the pool of darkness in the hall below but the hush told me that they were all in bed.

In my stockinged feet I crept to the scullery, a

lantern in one hand, and found the keys to Gulls Cry. Outside, I made my way to the path, keeping to the shadow cast by the house. On the beach I turned to look once at Iskar, but the building was in darkness. The sea lapped against the shoreline and too soon, I was standing at the door, my teeth chattering with anxiety and trying to turn the key with trembling fingers.

Inside, it was pitch-dark; decay and rot smouldered in my nostrils. My chest was tight as I put the lantern at my feet and fumbled for matches. It was deathly quiet and the strike loud from my match case, then the lamp bloomed into a globe of sulphurous light. My gaze drifted up the staircase to the corridors above. The eaves groaned and I summoned my courage and placed my hand on the banister and climbed to the first landing.

At the second set of stairs, I stopped and caught my breath. I did not like the way the lantern flung its light, creating a blinder darkness behind it. I ascended until I reached the top and went into the lumber room. This was where she had taken me in the dream and I felt in my heart that this was where I should be.

I was conscious immediately of the isolation, the chillier atmosphere. The dirty windows gave no moonlight and I tried not to imagine ghosts. Swinging the lantern, I saw the piles of furniture and fabrics – a cracked washstand and basin. I examined everything but there was nothing to hint at undiscovered secrets. Finally, I was left only with a pile of trunks.

For a heartbeat, I regarded their bulk and thought about leaving. But I recalled the fury of Hettie's eyes in the dream and I took them down, unclasped their rusty latches and disgorged the contents. Most were empty but for dust and beetles and I felt the beginnings of failure.

Soon my knees ached from the pressure of the boards and sweat flowered on my brow. I began to believe the task was fruitless.

Only one case remained and I opened the lid wearily. Unlike the others, this was not empty, but full of clothes – a mismatch of faded frocks and darned stockings, underskirts and grubby corsets. The sour, nettle-smell of stale sweat still clung to the fabric. I was discarding a blouse when something about it snagged my attention. It had a round lace collar and then I discerned it, so faintly and so overladen with other scents that it was barely recognizable – something flowery. And I knew two things: that I had seen a blouse like this in the photograph of Hettie, and the aroma in the air was hers.

I emptied more contents – underclothes, aprons, a comb set and handkerchiefs.

Holding the lantern closer I examined the collar

of a dress and found a label stitched there, went to another and another; all bore her name – these items belonged to Hettie. Why would she have abandoned them? Even if she had gone in haste, would she leave so much, and why was her case hidden?

I studied the trunk itself and found another compartment containing a heavy purse. I freed the clasp and there inked over the grimy fabric was her name. It was weighty with coins. Too weighty. And my heart began to slam against my chest. Hettie had not been rich; this must surely have been all the money she owned. She would not have discarded this if she had fled voluntarily.

Finally, my fingers touched upon a cloth bag with a jumble of knitting. Taking the items from the compartment, I hurried to repack the case and returned the trunks to the order in which I had found them. And all the while a sick sort of knowledge grew belly upwards.

Finally, I paused in shock. It was clear. There could be no mistake. Hettie had never caught the boat to America.

She had never left Skelthsea because she had been dead before that, and I recalled again her presence and the sound of her voice outside my room and the rage that had leaked from her dead eyes. I understood that fury now – she had not died on the boat. She had been killed here on Skelthsea and someone had attempted to conceal it by hiding her case at Gulls Cry.

Hettie had been murdered.

37

I could not leave the house quickly enough and it was only as I met the path to Iskar that I turned to view Gulls Cry one final time. The wind had picked up and the tide was higher on the sand. I thought of Hettie, not just as the witch she had been, but also as a woman whose life had been stolen. For surely, if she had died by accident, what motive could there have been to conceal it?

It was quiet in my room and I sat heavily in the chair, giving my heart time to cease its rattle. Hettie had been murdered. That was the unpalatable truth that had kept out of reach, and her face remoulded itself for me. In that moment, I knew something else – something that I had not been completely sure of but must now certainly be true: that Ailsa had most likely spoken the truth when she had told me that William had been killed. Before, I had not been able to justify such an abominable act, but a possible reason was finally evident – what if William had witnessed Hettie's killer? What if he had told? And what about Mary's muteness? Did she know too? All these weeks,

I had wondered what powerful key kept her lips so tightly shut. What logic could be more potent than the knowledge of her brother's murder to hold her to silence?

My hands shook. All wrong. All wrong. I examined the purse again – the coins that she had saved and would never spend. There was little in her sewing bag – needles and two knitted garments – a baby's tiny bedjacket and hat. Even with the gravity of horror, I could see the care which had been taken to produce them. I imagined her then sitting by the fire, her cheek caught in the lamplight, her clever fingers on the wool.

I had thought that, by banishing her ghost, I had made Mary safe. That could not be true any more. The logic no longer held, because Hettie had been killed not by a ghost but by human hands, and probably William too. If Hettie returned from the grave later and tried to harm us – it was because Greer wished it. I recalled again that blaze of love on Greer's face as she looked at Paterson. I felt the fierceness of Greer's passionate nature – saw her hands taking Hettie's life and then that of the boy who saw her do it.

It came to me then with a jolt – something that I had failed to connect before. Hettie had first hummed her lullaby on the second day of my arrival – some weeks before her body washed up. Whoever summoned her had known then that she was dead, before everyone else learned the truth. And therefore they were either the one to have committed the murder or to have witnessed it.

It had to be Greer – Greer who wanted the rival for Paterson's affection out of the way. She was more rotten than I had realized – far more so.

My thoughts weaved sickly. I hid what I had found in one of the unused rooms and stood at the window. The island was still. 'Stay away from the dolphin,' Mary had written, and its meaning was chilling; I must not discover what had happened to Hettie as William had. I should have stayed away, remained in ignorance – to have remained in the shadow of the truth would have kept me safe, but I now knew it all. Knew everything.

I crept along the corridor and to Mary's room; her skin was as pale as the inside of an eggshell and I saw finally what I had been blinded to for the past days, blinded by the demonstration of her affection and my resolute belief that she was healing. I had only seen what I had wanted to see. I had not read her truly – because, beyond the thawing of her heart, she remained haunted. Hettie and William's going had not eased Mary – her condition had worsened and the lines of anxiety were deeper, her skin more taut. Whatever preyed upon her fears remained.

I returned to my room, to the cold hearth, and

watched the candle flame as if it could save me from the dread that pressed once more against my chest. I remembered those nights that Hettie came and her lullaby replayed through my head – the way her rage had cut the air. I need no longer wonder at it. I closed my eyes and saw the dream where she lay upon the bed as still as a doll. The bloody tear in her dress. I recalled Mrs Lenister's words when Hettie had the newborn in her arms and sang – *she held it like her own*, she had said.

The room seemed to pall, the air to grow more chilled. From the windows cloud yellowed in the moon's light. I saw again her reflection in the mirror – pain breaking across her face. Lamplight on her cheek as she stitched the tiny clothes.

And I knew. The knowledge left me faint – Hettie had been with Paterson's child. I felt it as a truth. For a moment my emotions stumbled into pity and then I recalled how close Mary had come to death and I hardened my heart.

Greer had murdered not just a mother but her unborn baby as well.

Dawn came, uncaring of the truth. I rose from the sheets heavier in my bones than I had been the morning before. If Mary saw the change in me she did not show it, and the bright sun drew us out again to the sand and sea.

The day seemed too dazzling not to banish the weight of horror, but every rock, every tree was painted in the same fearful knowledge, and the dread that I believed I had shuffled away wound tighter and tighter once again.

That night, in spite of exhaustion, sleep eluded me. My head ached. I did not know how I could manage the return of fear – fear made worse by what I now understood. I began to feel again the dissembling of my senses. An hour I spent, turning on the sheets in restless thought, and at the end of it, I had to know the truth and so I went to Mary's room and shook her awake. She did not sleep deeply and her eyes came open quickly and gazed at me with alarm.

'Mary,' I said. 'You told me about the dolphin and to stay away.'

She sat up, the pulse beating hard at her throat. 'Mary,' I pleaded, 'if you know how Hettie and William really died or you are in danger, you can tell me.' I put my hand on hers. It was nearly on my lips to reveal what had happened the night Ailsa and I had sent Hettie and William back to the grave. 'Don't you know, Mary, what lengths I would go to to protect you?'

My throat tightened. 'You are like a sister to me now. Like Clara.'

A look of hope sprang to her expression but before I could fully appreciate it, it was sunk in fear.

'Who?' I begged. 'Who means you harm?' Because I was convinced that someone did. 'It's too late now, Mary. Do you understand? Too late to keep this terrible secret. They were murdered, weren't they?'

Her eyes grew wider.

'Yes,' I said, 'I know it all.'

Her skin paled and she reached for Bobbity, clutching her to her chest.

'Mary,' I took her face in my hands, 'Mary, you have been silent long enough.'

Her mouth worked. Her lips paled but then her eyes found mine and fixed there.

'I can't,' she mouthed in anguish, 'I can't.'

'You must do it, Mary. You can speak safely to me. Greer, was it her?'

I held her hand tight to mine and something crumbled behind her look. She leaned in close to my ear and finally, finally unglued her tongue and, in a voice so soft I had to draw even closer, she told me everything.

38

Mary explained how William had confessed to her that he had seen Hettie's body dragged from the beach, wrapped in sailcloth and taken out to sea. He was afraid that he had been witnessed but refused to tell Mary who had been in the boat. He knew the risk, even then. Although William had not divulged the identity of the culprit, she had understood his fear. Not more than two weeks later, his body had been found on the rocks. She had guessed he had been murdered and could think of no other way of protecting herself except by not speaking. I wanted to weep at this – at the childish trust that holding back her words might keep her safe. The weeks had passed and she had held on to her silence and then I had come and begun to ask questions and Mary had found a note on her pillow.

William died because he could not keep his tongue still. If you speak of what you know I shall still not only yours but Elspeth's too.

The weight of her burden was so heavy on my heart that I thought it would break. She threw herself into my arms and wept, and in her tears I felt the grief and fear that had haunted her since the day of my arrival.

When she had finished, the silence grew around us. 'What will you do, Elspeth?' she whispered.

I did not know what I could do. But I looked down and stroked her cheek. 'I will think of something. And your voice, it is a lovely thing, Mary. So many would be pleased to hear it.'

'I'm afraid,' she said again. 'I am too afraid to speak to anyone but you.'

I understood.

'Will you stay here tonight?'

I climbed in beside her. No more questions remained and it did not matter whether or not she knew who wrote the note; what mattered was they believed that she did. I thought of Greer's pen inscribing the paper – printing in ink a threat to the child who had once been in her care – and was disgusted.

As I held her warm body to mine, I felt the slipping of hope. I nearly gathered up all the evidence that I held to take to Miss Gillies, but she would not want to believe Greer responsible, and what would she do?

As if sensing my thoughts, Mary pulled away. 'You must not say anything.'

She was right, I must not. So I lay next to her and watched the shadows play against the walls and listened to the endless exhalation of the sea.

The next day the cold settled damply on everything it touched.

Soon the rain began, slipping on the windows and clinking into a pail by the door, keeping myself and Mary in. When the time came for her lessons, although it was too wet to leave, that is what I did, needing to feel the fresh air on my skin. I must think.

My feet slipped constantly on the wet pebbles and soon my face and coat were slick with moisture. In contrast to my usual experience, the further I walked from Iskar, the greater my anxiety grew. I tried to think of all the ways we might escape together, but it seemed like an impossible thing. Droplets stung my eyes and seeped into my cuffs and collar. All the while my heart was tight in my chest. Fear had found again all those places I thought had been closed after the banishing of ghosts.

I had walked nearly as far as Gulls Cry when I heard it – a high, childish scream ringing out over the wind. Terror gripped me. I imagined Mary dragged across the shingle to the water's edge, imagined her in Greer's clasp. Even as the scream died on the air, I was racing back across the shingle, a pain in my chest, my legs straining and feet bruised by the pressure of stones. And then I was in Iskar's hall, panic making me shout Mary's name.

Miss Gillies came out of the schoolroom. 'What on earth is this? What's the matter, Elspeth?'

I could not speak because my breath was trapped in my lungs. Behind her stood Mary, her eyes wide with surprise and alarm as I dripped water on to the tiles. She had not screamed at all, and if she had, I could not have heard her from the beach. It must have been the call of a bird – or something terror had pulled from my imagination.

'I'm sorry,' I said, 'sorry, it's nothing. Forgive me.' Miss Gillies' expression grew icy and Greer came into the hall and watched me with a smile.

Without even removing my coat, I ran up to my room. The rest of the afternoon passed with a strange unreality as if I had left my own skin, and my limbs walked and mouth talked but I was no longer their master and I watched myself unpinning and brushing out the tangle of hair, grains of sand falling with a scuttle of sound upon the dressing table.

When lessons came the next day, I made my way along the valley, my steps laboured. I had no desire to visit Bridget but I had promised; what I knew sat too stone-like on my chest.

She beckoned me in and bade me sit in the familiar chair and organized tea.

I answered her questions and all the time the secrets pressed.

Soon, she put down her cup, her face grave.

'My friend, something troubles you. I can see. I have watched you, seen at times that you are unhappy or concerned, but I do not think that I have ever seen you cast down to this level. You are pale and there is fear in your eyes – fear and pain.' She reached both hands across and clasped mine. 'Let me help you. Tell me what has happened.'

'I hardly dare say,' I said finally.

'You must. I will not think you silly and I will not take what you say back to Miss Gillies. Is it Mary?'

I shook my head, the burden of knowledge swollen in my throat, and I could not stop myself; the barrier broke and I began to weep. Once I had begun, I did not cease until the tears had run their course and all the time Bridget held me. I remembered then my mother's arms, my cheek against her chest. Those days had been too few. She had died giving life to Clara. One heart for another.

'Listen to me. You cannot keep whatever this is to yourself; it is clearly something of importance.' She took my wet face in her palms and I lifted my eyes to hers. 'I could never think you foolish and I promise that whatever you tell me, whatever I feel about it, I shall keep in confidence.'

I believed her and, as Mary had done not two days past, I told her everything.

When I had finished, shock rendered her cheeks pale. We sat in silence as she absorbed what she had learned.

Her eyes were pained. 'It was well known that Hettie and Paterson were lovers. They hardly hid it. Stupid girl. And Greer,' she looked at me with concern, 'her feelings for Paterson were gossiped about long before he took Hettie. It had not occurred to me how that might have affected how she felt about her friend.' She frowned. 'Elspeth, it is not good that you have crossed her. I believe her capable of anything. As for Paterson,' she almost spat on his name, 'he is a bad creature if there ever was one. That wife and those children see his fists more often than they go to church. Perhaps he too had a hand in it. He would not have liked her to be with his child. We must notify the Edinburgh police. Justice must be served.'

'It's too late for police. If it becomes known, do you suppose that Mary will make it alive?'

'You must tell Miss Gillies.'

I shook my head. 'Miss Gillies will not want to listen. You know too well that she does not care for Mary and she owes a debt to Greer because of her mother. In a few weeks she may well have Mary committed to an asylum, anyway.'

'And she would be safer there,' Bridget said. 'Dear God. You must leave, then, my friend. You must not risk yourself. Go on the next boat. It would be madness to stay.'

'I cannot leave Mary.'

Tears sprang to her eyes, 'You are a better soul than Mary could ever have wished for.'

'I have thought of little else but leaving with Mary, yet how can it be achieved without someone knowing?'

She was silent for some time then she turned to me. 'It has to be possible. It must be, Elspeth. My sister lives on one of the neighbouring islands and could secure passage to the mainland. I will give you a letter. And there is a fisherman here who owes me a favour. He would not break his silence if I asked him to take you. I believe that he would. He could.' She clutched my palm.

'Are you certain?'

'It's dangerous but you have no other choice. You have money? I have spare if you need it.'

'I have enough and sufficient to start again somewhere. We could go to London. I would find work.'

'But if you were caught, you could go to prison,' she said.

I smiled because I knew by then, without a doubt,

that I would risk my life or imprisonment many times over to save Mary's.

Finally, she said, 'You're a good girl, Elspeth. I shall miss you.' Her composure broke and I thought she might cry.

'I shall take everything I have of evidence to support my case should I be caught.'

'I will do my part and let you know as soon as I have news.'

I rose. 'I must get back. I don't know how to thank you, Bridget.'

But she seemed lost in reflection and gave me only a small smile. I reached out and took her hand. 'You have been a good friend to me. The best,' I said.

Her eyes were sad. 'I believe, in time, we would have become great companions, but it was not to be. I will miss you so much, dearest Elspeth.'

39

Outside, I paused. The valley was sculpted with the evening's soft light. Iskar rose into the sky and I regarded it as if it were an enemy already vanquished. As clouds shifted and the setting sun glanced along its roofs and windows, I was aware of it as an entity, as if it were part of the island itself, a growth from the soil. Below it lay the sea, a silver ribbon cut with rock.

The house was still when I entered; the clock in the hall turned gratingly on the hour, as it had from that first day, and the portrait of the twins dulled in the lamplight. I stepped closer and searched William's face. Whatever his sins, he had not deserved the fate that had awaited him.

I tried to imagine what it would be like once Mary and I had gone. I could see Miss Gillies ordering Angus to take the portrait down, and later, clearing out the playroom and bedrooms. Before long nothing would be left of them and Iskar would return to its ticking loneliness – the wind and the soft creak of wood breaking upon its silences and Miss Gillies sitting alone in the drawing room by the heat of the fire, a whisky in her hands, growing older as Iskar grew older about her.

That evening, we dined with Miss Gillies.

We ate mostly in silence, the cold curling on our ankles, but I was transfixed with Miss Gillies as she drank her wine, the way the dent at the base of her throat moved when she swallowed, how her fingers curved on the glass and how, every now and then, her gaze slid to Mary and dislike came down upon her eye like the third membrane over the cornea of a bird.

And I knew that I had been right – that Mary's aunt, had I told her what I knew, would not have troubled herself to listen or pay heed.

In the silences, I tried to imagine London, the sound of carriage wheels and the call of flower girls. In my head I calculated my funds, mapped our journey, and I think part of me had already left Skelthsea.

In bed, my thoughts drifted back to Swan House, a place that seemed so distant now as to be a dream. I remembered the smell of the meadow clotted with buttercup and cuckoo flower, and heard the saw of crickets in the grass.

It hurt, but I did not resist and allowed the memories to roam as they pleased. I recalled the minutiae then – the chip in the sideboard where Barbara had dropped a jug, the photograph of my parents on a punt in Cambridge, and I knew that if I did not keep remembering, the day would come when I would forget. Already, it seemed that those easy days had slipped behind a door that could never be re-opened. My heart was here, with Mary who lived and breathed.

The morning brought the beginning of strong weather; wind buffeted the panes and beyond the glass the sea tore at the edges of the island. After a brief walk that left us shivering, we returned to an empty house. The servants were helping Miss Gillies with her quarterly tenant visits, armed with food parcels, and I was grateful not to have the scrutiny of others – surely my nervous excitement would be there for all to see. I suggested a game of hide-and-seek.

Mary went off first and I sat in the hall chair, remembering Miss Gillies' story, and reaching down, found the loose thread and pulled at it with my fingers as she had done. I tried to imagine Miss Gillies as a child, her face unmarred, her future unspoiled. With my ear I traced Mary's step and when I had reflected enough, I went to find her.

When we had finished playing, we raced through the corridors, uncaring for once of decorum and noise, and ended up at the kitchen table where I made tea and found cake and meat that Mrs Lenister had left. The room was warm and sweet and I felt such regret that we had to leave; I could have found a home here, amongst the decaying grandeur, and I recalled that first day as I stood atop the ridge with all the beauty of Skelthsea falling before me.

'What now?' I asked and Mary grinned, keen for another game of high activity, but I became aware of the lightness at my neck and realized, with dismay, that my locket was no longer there.

'My locket. I've lost it.'

Mary looked at me with such alarm that I would have laughed had I not felt so near to crying, and I tried to think of all the places we had hidden and all the dusty corners in which I had crouched.

'Mary, if I take the upstairs, would you check the ground floor?' She nodded, running off without hesitation, and I retraced my steps, scouring the carpets with growing distress. So many rooms, so many places. Eventually I descended the stairs and found Mary coming in from the scullery corridor.

'I cannot find it,' I said, and she gave me an anxious look but was not quick enough to hide the nervous clutching of her fingers as she put one arm behind her back.

'What have you there?'

She did not answer.

'What is it?'

And slowly, she brought her hand round and opened it. There on her palm lay the locket. Jubilation was quickly followed by confusion at her apparent reluctance to show me.

'Why didn't you find me right away?' I took it and peered down. Mud was engrained in the silverwork and in the chain as though it had been trodden into dirt. I looked at Mary with wonder and took in for the first time the wet tendrils of her hair and the cold flush on her skin. She had not found this in the house. I realized then that I could have as easily dropped it outside Iskar as in. But that was not where I had suggested she look.

'This was outside, Mary,' I said. 'You found this outside.' I gazed at her with wonder. Her expression was too guarded. What had I missed? When understanding came to me it left me breathless. She had known by some strange talent. And I recalled what Miss Gillies had said of Hettie's ability to find lost things. Perhaps it had never been Hettie's gift at all but had always been Mary's. I remembered Mary's curious skill in reading my thoughts. Mary – it had never been otherwise. I was sure of it. But before I could say anything her feet were thumping up the stairs followed by the distant banging of a door.

40

I stood for a moment, shock pinning me to the tiles, and then I ran after her. It did not take long to find her: she was sitting on her bed, her back to me.

I sat beside her with a sort of horrified awe. 'It was always you, wasn't it? It was never Hettie that had the gift?'

She did not reply.

'Hettie covered for you. Why did she do that? Were you with Hettie the day Greer's necklace was found? Was it you that found it?'

Her look was half sad, half angry. She opened her mouth, checked around to make sure no one was there to hear. 'Yes, it was me. William said I should not tell, and Hettie liked what people thought of her. She did not care if they said she was strange.'

We sat for a while with only the sound of the wind and the lap of the tide from beyond the windows. As I scoured her face, it seemed to me to be one of someone many years beyond her age. I had underestimated her. The evening that she had begged me not to leave her – I had been so distracted that I had misunderstood her and replied that I had not been planning to leave Skelthsea – that was not what she had been asking. She had been asking me to stay that night with her, in her room. She knew that something would happen. And then Hettie had taken her to the Fiaclach. Finally, I recalled the time she told me that she was to die.

'Are they always true?' I asked. 'Your predictions?'

Her eyes were as clear as glass: the answer was there.

'Can things be changed?'

She picked at Bobbity's hair. 'I don't know.'

A sudden squall of rain snapped at the panes and somewhere far off a tile rattled. Her fingers tightened on the doll. Below us, the hall clock chimed the hour and then, as though spent, the house fell again to silence, but some sense of safety and confidence that I had been clinging to slipped a little away.

I got up and went to her basin and rinsed the locket, drying it on the hand towel, and then I clipped it once again around my neck. I saw my face in the mirror but did not want to meet my own thoughts. Behind me, Mary looked up and caught my eye and I could not, as hard as I tried, read her.

The remainder of the afternoon we spent quietly, but she was unhappy. She had not wanted me to know. Was she afraid that I would be disappointed?

Did she think that I would tell someone? Why would she not trust me if she trusted Hettie?

Slowly, the wind rose around the house and the trees shook, making shadows on the faded wallpaper. Miss Gillies returned, weary and pale, and fell asleep with a whisky on the tray before her. Mrs Lenister brought broth and oat cakes to the playroom but I could not eat, and although I tried for conversation, it seemed that every word I uttered cut deeper into the silence between us.

Bed could not come too soon and I lay listening to the wind pulling at the island. Wrong, wrong, wrong. I sensed it again, something bent in my logic, something I had missed, and I felt as if I were falling – falling through water, through the unseeing cold, and missing some truth that would not show itself. I had lost the voice tinged with pipe smoke, lost the sound of Clara's laugh. All I had left was the fragile stem of my own bones and my love for Mary. But that too left me cold with doubt.

The storm came and left in its wake a sort of unearthly stillness, as if the tide itself was holding its breath before breaking loose again upon the shore.

The following morning, the fires burned downstairs and the scents of Iskar seemed new – the breath of paraffin and wax, of peat and cooking, and behind that, always, always the smell of the sea and its clinging brine. It was absorbed in everything – in the fabric of tree and grass and rock. To live here was always to be somehow part of it.

At breakfast, Miss Gillies chatted inconsequentially, her mood lifted by an onerous task discharged, and Mary sat more quietly than ever, although we caught each other once in a sidelong glance and her eyes found my locket.

It was a strange walk that we took that morning, across a beach littered with broken branches, pieces of fence, a twisted lobster pot. The sun was hidden, leaving no shadows on the sand. Later we made our procession to church, Miss Gillies in front, Mary at her side, and I could not take my eyes from their stiff figures: aunt and niece, the last of the Gillies line. Halfway up the path, Mary turned and regarded me with a flat look.

At the village, there was a pause and the usual flurry of conversation, but instead I turned and viewed Iskar: its gabled roof, arabesque arches and then higher up the tiny panes of the attics. I felt like a giant then – as if my fingers had prised open the house to examine the contents, to move furniture from this room to that and to pick up the people – Hettie, Greer and Mary – as if they were dolls.

I heard nothing of the service or the echoing hymn

and earnest sermon. Afterwards, I waited in line behind Miss Gillies as she made her goodbyes to the Argylles with the half-light of dusk beyond the doors. Even before Miss Gillies had finished, I felt Bridget's eyes upon me and I pricked awake. Mary turned and watched me with such intensity that I believed she knew what was in my thoughts.

As Argylle clasped my hand and muttered pleasant good wishes, I could feel the urgency of Bridget's mood, and when she took my palm the quickest of glances showed me that she had passed a folded note into my fingers and I closed my glove around it and slipped it into my pocket. Beyond the clatter of voices the island was still. The weather vane barely turned. My heart began to beat hard in my chest.

Back at the house, I made my excuses and took the stairs to my room and pulled out the note with trembling fingers:

Be at the North Beach tonight at 11. Bring only what you cannot leave and money for the boat. Dress warm and take care.

Your friend

Tears pricked my eyes and I tried to envisage our escape and the London I had been imagining with its boarding house. I pictured the plump-breasted owner and narrow rooms – a press with a faulty hinge – tables laid for breakfast and dull polished silver, cheap candlesticks and outside the rattle of wheels on the cobbled streets. I saw us in twin beds in a whitewashed room. Saw Mary's open expression, free finally of the ghosts and fear that had followed her on Skelthsea. I would teach her all she needed to know to lead a useful life. I would be the sister I had been to Clara. I would be all those things.

My mind turned over what could be kept and what could be left: I had to consider Mary too, but I was not so poor that I could not buy what we needed in London, and had enough for some months in cheap lodgings until I found a position. Yes, it was all possible.

And once free, Mary would have the confidence to speak to all who addressed her, and as for her abilities, hadn't I heard that children grew out of their peculiar sensitivities with age? This strange gift that she had, might she not leave it behind with the years? And would Miss Gillies even seek for her?

Dinner was taken with Miss Gillies but my mind was so busy that I was hardly aware of that final meal until Mrs Lenister came in with her smile and placed a bowl upon the sideboard. The candle flames bowed to her movement. She smelt of herbs and something sweet. Miss Gillies turned her cuff and looked up. 'Is that the last of the plums?'

'Good crop, but that's the lot until next year.'

'How quickly winter always comes,' Miss Gillies said. 'Before you realize.'

'Aye, that's the truth, but spring arrives as quickly too. I made the sponge you like for the fruit.' A warmth passed between them, so fast it could easily have taken place unnoticed, and I realized then all the kinder notes inside the phrases that I had missed.

When I put Mary to bed, the words about our escape were on my tongue but I found that I could not utter them. I would wait until the moment we left. And if she already knew, then she could ask me, I reasoned.

As I sat later with Miss Gillies, I allowed myself to drift in the quiet of the room, in the fire's flickering shadow and the tick of the clock. I found myself looking with a different eye upon the birds in their case and the porcelain vases, and I realized that I was committing them to memory and that one day, long into the future and with enough distance, I might relive the time that I had spent here, might even dream that I was sitting upon this very chair with this very book I had with me now.

It grew late and I stood; Miss Gillies had fallen

asleep. I walked to the door, but as I touched the handle she stirred and woke, and I turned.

She tried to mask a yawn. 'Is that the time already? I must go to bed too. Goodnight, Elspeth,' and there was a pause behind her words and my thoughts softened. It was a final goodbye.

41

Upstairs, there was no time to ponder my decision. It was too late now and I must see it to the very end. I pulled out a chair and took a case from the top of the press and laid it on the bed. It did not take me long to pack – I had so little that could not be easily replaced. I waited, listening, and then I crept to Mary's room and took the clothes and the toys that she most valued. And even when I had finished, the case was but three-quarters full.

The air smelt of damp satin and leather and I closed the clasp and waited, reminded of that day that I had stood on the pier at Mallaig and watched the boat slice the waters. I caught a glimpse of myself in the mirror. I was not the person I had been then. I was finally fully grown.

I recalled that last morning with Clara, how the wind had carried the spring's scents through the open window and into the parlour. How we had planned our new dresses and counted how many coins we would have spare for some good meat. I had turned to pick up my pencil and continue my list when Clara had placed her hand over mine.

'Thank you,' she said and there had been a look of such earnest gratitude that my eyes had filled.

'Don't be silly, Clara, it is only what you would do for me.'

'I would do anything for you, Elspeth, if I could,' and she reached into her pocket and passed me something, clumsily swaddled in brown paper.

'What is this?' I said.

Her love laid itself over my heart and my eyes burned as I fumbled with the wrapping.

'So that we shall always be together.'

I opened the locket and there, forever held in silver, were our two images.

When it was time, I tiptoed into the hall where the dusky smell of lamps bloomed on the air and I woke Mary from her bed. She asked no questions but let me help her dress and it was only as I handed her Bobbity and she noticed my case by the door that I told her of what was planned.

Our feet were loud on the pebbles, and as we rounded the bay the smell of pine hung in the air like a final song. Somewhere above us an owl's call shivered through the night. As we approached the north beaches, cloud began to pass across the moon and I crouched in the shadow of the cliffs and lit our lantern. The tide was coming up and licking at the silty sand, but there was little noise save our boots and laboured breath. Mary, who had hold of my hand, stopped, pulling me sharply to a halt.

'Hurry,' I said, 'we don't have time.'

'I'm scared,' she said.

I put down my case and bent and kissed her forehead. 'I'm scared too.'

We had reached the end of the strand and were climbing on the rocky plateau towards the caves. Above us, gulls fluttered and cawed gently from their roosts. My arm grew sore and I swapped the case for the other hand. Finally, we were on shingle once more and Bridget was ahead, wrapped in a scarf and heavy coat. My heart let go of its fears and I ran to her.

'Quick,' she said. 'Hurry,' and I followed her to the little boat pulled half up on the sand.

'Take care here, take care. I had to row it round, so as not to be seen. The bigger boat is behind the point.'

I glanced at Mary, at her liquid eyes. 'It'll be all right, Mary. Soon we'll be safe.'

Bridget took my case and helped Mary into the boat and then together we pushed it into the waves and waded in until it was afloat. I gasped as the water found my skin and my skirts billowed up like an umbrella.

'Quickly,' Bridget said. The vessel tipped a little but Bridget steadied us with the paddles.

I looked to Mary, clutched her hand and gazed up at the stars, scattered like white dust across the heavens. My feet and hands were numb but excitement took away the feeling of cold. Bridget began to row and I turned to watch the beach retreating.

We did not speak. I was aware of the steady cut and splash of the oars as they cleaved the waves. The air grew wetter. The boat swung, leaving a white rip upon the ocean. My feelings were so bright and highly charged that I could not separate apprehension from hope, excitement from fear. We were heading towards the open ocean now where a sea fret began to creep in low over the water.

'The mist,' I said. 'Will we be all right to get to the other island?'

She gave a low chuckle. 'You're still a city girl, Elspeth. That's nothing.'

'How will we get on the other boat?'

'He's anchored off the water yonder. He'll throw me a line and you and Mary will have to ladder it up the side.'

I thought incongruously of the civilized afternoon teas at Swan House: I could never have imagined such adventures would be ahead. But suddenly, something in the air changed, an alteration in energy. Bridget's scrutiny was fixed seawards but Mary had gone still. She took my gaze and dragged it to a little panel on the boat's inside and my heart fluttered.

The Dolphin. The boat was named 'The Dolphin'. I looked at Mary again. It meant nothing, a coincidence, but Mary's eyes held a knowledge that I was not party to. Mary, with her gift. *Don't go near the dolphin*, she had written. All this time I had been fixed on the brooch, but what if that was not what she had been talking of? Mary's eyes slid to Bridget. Was I to believe that Bridget, my friend, my one companion on Skelthsea, was the person pitched against us? Not her. I sought out Mary's look and she answered it. And with the dawning of horror, I knew that Bridget had always been the one of whom I should be afraid.

I turned slowly to her, the comprehension chilling, but it was too late. She had observed what I had been too late to hide. She stilled her oars, leaving just a splash at the bow.

Silence, not even the wind, only a yawning quiet.

'You,' my voice was ragged.

'I killed Hettie, yes.'

I twisted to look at the shore but it was too far away. I did not even know if Mary could swim.

'Why?'

'I would have put up with an affair, but she was going to give Robert what he most wanted – a child.' And behind the kindly lines of her face, another showed, cut with grief and bitterness.

'I thought it was Paterson,' I said.

'She could have had any man she wanted – she may have slept with Paterson but it was Robert who lost his heart and gave her what I wanted with all mine.'

'And William saw you?'

'They were evil. Skelthsea was better without them.' She took up the oars again.

'Let us go,' I said. 'Take us to the boat. I won't tell. I'll never tell.'

The slip of the wood through the water was her only answer and a new terror clutched at me.

'Tell me that there is a boat,' I said.

'No boat. This is it.'

'I cannot believe this of you. Please.'

But suddenly Mary stood, rocking the vessel, her face creased with fury.

'They were never evil. He was not bad. She's lying, Elspeth. Hettie was good and William was good. Hettie protected us. She made it up.'

Bridget's lips tightened. 'Hettie was a whore!'

'You made them look bad. William never hurt a creature – never in his life. He loved them. And Hettie, Hettie was kind.' Tears began to fill her eyes and I turned to Bridget in horror and disgust.

'You did those things? You made it so that

everyone thought it was William and Hettie.' And I was assailed with a sudden hopelessness. I thought of all the conversations she had fed with lies until I was so full of them it left no space for truth. If Bridget was capable of that then there could surely be no part of her that I could reach. I pulled Mary to my side.

Bridget shrugged and I was mesmerized by the strong rhythm of her shoulders and how the boat went relentlessly on.

'I thought,' she said finally, 'if I made it so that she was wicked, that Robert would leave her alone. Although his flesh is weak his moral spirit is strong. He could not have bedded a witch.'

I felt sick. 'Stop now,' I pleaded. 'Take us back. Haven't you caused enough pain, Bridget? I thought you were my friend.'

'And so I am.'

But the blank stare she returned showed me the truth. She did not care. And then behind us there was a shout. I turned. A figure scrambled across the rocks, hair loosed down her back. 'Stop,' she yelled, 'I won't lie for you any longer.'

And I realized, with shock, that it was Greer.

She stood, her face white in the semi-darkness, clutching her hands as if in despair and a spark of hope lit my chest.

But in that second, with my back to Bridget, there

was a push and I pitched forward and was in the water, the cold a bolt to my heart. I reached up for the boat to cling to but only for a moment was I on the surface. Mary's scream tore at the air and then I was dragged under, sinking through the blackness, my body weighted like stone.

42

Time slowed and I was bound by the water – a blind creature in an icy womb. And the world paused. The drag from the bottom pulled at me and I imagined the dolls in their coffins, imagined Hettie – her hair waving in the water. I had been wrong; Hettie had never been bad, and in that moment with my life shrinking I understood. Hettie had not taken Mary to the Fiaclach to harm her but led me there to save her. In everything, she sought only to protect. And I had sent her away.

The silence was as deep and still as distant galaxies. Every piece of my life came polished to diamond sharpness, fragments hurled at me with the speed of comets: the coiling smoke of Swan House, my mother's face with death upon it, the warmth of Clara's hand – no regret: my heart was as flat as paper.

And then, through the darkness, through the myriad of images, I heard her. 'Elspeth,' she said, and the letters were not jumbled as they had been in life. Her voice was as clear as birdsong. For an instant, she was there – her beautiful, imperfect face – Clara. I not only heard and saw her but I felt her, once again, like a piece of my soul and I came awake and began to struggle – a slow, weightless dance that dragged me from death, while above, the moon pierced the dark like a silver shilling. I fought until I came gasping and coughing to the surface. The beach was too far but there was Greer, on a lip of rock that jutted into the sea.

'This way,' she screamed.

I whipped my head round and saw in alarm that the boat was turning behind the cliffs, Mary's ashen face frozen in horror. I pushed through the waves, my arms aching, to where Greer lay on her belly with her hands outstretched. The tide pushed me closer and finally, I felt the grip of her palm. For a moment, our eyes met, then she dragged me from the water, ripping my stockings and skirts.

I had hardly stood when Greer was running ahead along the narrow ledge that abutted the cliffs and I followed, pulling off my sodden boots and skirts as I went.

And then we were high up, above a sheer drop where the boat below was as insubstantial as a toy. The mist was taking hold, thickening like a web around them. Wind tore at my hair and whipped through the crags.

'Bridget,' I screamed, but Greer turned to me and shook her head. Her hand went to mine and passed something. Looking down, I saw in my palm what she had given me and I gazed at her wonderingly.

Above us the grasses whistled and the moon dipped out of cloud and I was aware of a strange, otherworldly certainty about what I was to do. Slowly, I lifted it to my lips. Its texture was soft, and as it lay upon my mouth, seemed to adjust to my own skin. For a pause, the edges of the world's limits stretched and shivered on the night's breath.

I blew.

The sound was familiar, thin and reedy, but this time, seemed to come from deep inside me as if the widows' whistle were a mere appendage and the calling back of Hettie was a plea from my own soul. I felt the essence of her on my flesh, the scent of her in my nostrils, and the imprint of her heart laid itself over mine. In that moment, I saw her inside out – the flaws and the strengths. I experienced the tearing of her loss and the burning of her love. She had not been wicked, only human. Only human. Like me.

Greer began to scramble down the rocks.

I followed her down. We were near the water now and the boat, disrobed momentarily from the fog, was revealed – Bridget at the bow, Mary at the other end.

'Mary,' I made to jump but Greer gripped me hard and pulled me back.

'Don't be a fool. You'll drown. Look –'

And as I watched, the mist swelled again and rolled like smoke until the boat was hidden.

'Mary,' I tried to launch myself once more.

'Don't be an idiot,' she spat. 'You'll die on the rocks first and it will all be for nothing.' She grasped my arm and we clambered further along.

Lower down, the silence deepened and I recalled the strange quietness of Hettie's lullaby, the sense of the physical world being twisted slightly out of kilter, the instinct that other forces had taken a hand. I felt it again.

There was a splash and my heart froze, and then the sound of the boat returning, but mist hugged the vessel and there was nothing to see.

'It's coming back,' Greer said, and we were running along the rocks and to the beach, the sound of oars growing louder. I dreaded what I might witness and I stood, terrified that only Bridget would return.

The shreds of cloud parted and for an instant, I believed there was a pair of slim arms upon the oars, but the sea fret fell again and then the vessel hit the lip of sand and we were wading in to drag it on to the beach. Mary sat alone.

She flung herself into me and buried her damp head to my chest. Wordless, I sank to my knees and held her until the fear had passed. I was too empty for talk, too shocked to speak, and when we rose, I could only put one foot in front of the other. Greer had already gone.

43

The hall at Iskar was still – peat smoke and dust. I was shivering. I sent Mary upstairs to change but I could not move. Greer came through from the scullery and placed my case upon the tiles. I had not seen her take it from the boat.

'I thought you hated me,' I said.

'Perhaps I do.'

Her face was waxy, her eyes masked, and she shrugged her shoulders and left. I climbed the stairs to get out of my wet things. Afterwards, I crept into Mary's room where she sat, draped in her counterpane.

'Are you cold?' I asked.

'A little.' Her cheeks were very pink.

'What happened?'

'Hettie came, Elspeth. She took Bridget into the water.' Her eyes were bright with shock.

'You loved Hettie.'

She nodded. 'But I love you better.'

'Let me get you something hot from the kitchen.'

She looked at me shyly. 'I'd rather you sat here.' So I sat beside her and put my hands over hers.

'Shall we tell Miss Gillies what happened?'

She shook her head.

'Did you make any of the pebbles?' I asked, curiously.

'Greer showed me. She told me they would help bring them back if I wanted them enough. They did come back, but then they left again and I did not use them after what you said.'

Mary believed that she had brought William and Hettie. At some point I would tell her the whole truth. Not now.

'Where's Bobbity?'

'She fell in the water.'

'I'm sorry.'

But she laid her head against me. 'It's all right, Elspeth.'

We sat for a while and then I stood. 'There's something I must do.'

In the kitchen, I heated water and found whisky and honey and made up a drink. The stairs to the servants' quarters were narrow. My candle flickered on the crumbling plaster and cold blew in from beneath doors.

At the top, I paused. Mrs Lenister's snoring drifted in the passageway. I tiptoed along the corridor to Greer, knocked and went in. She was sitting on the bed, the covers around her shoulders. Although she barely acknowledged me, she took the drink.

'How did you know that we were planning to leave tonight?'

'Mrs Argylle told me.'

I remembered what she had called from the rocks, that she would no longer lie. 'What has been your part, Greer?'

She stared at me, half angry, half sad. Then she leaned her chin on her palms and sighed.

'A long time ago, I took a confession to the minister, but Mrs Argylle heard all that I said to him.

'When Hettie began her affair, Mrs Argylle threatened to tell Miss Gillies everything she knew about me if I did not help her. That would have been the end for me.' She looked up with something like bitterness. 'I'm not like you. I do not have money or education or good looks. I do not have skills that I can take elsewhere. This is the only home I have ever known and the only place that would ever feel so. Miss Gillies is the one person in the world who cares for me. And so I did what Mrs Argylle asked.' Her voice fell to a whisper.

'She made me take the animals up to the stones and she spread the word of Hettie and William's wickedness. And when that failed she killed Hettie. I left the goodbye letter she wrote for Miss Gillies to find and hid Hettie's things at Gulls Cry. I did as she asked. She told me that William had seen. I begged him to keep his mouth shut, but it was in his eye whenever she was there. He was not the sort of boy to keep a secret. Not one like that, and Mrs Argylle knew it.' She gave something like a gasp. 'I could not believe that she would kill a child, though; I did not dream that she could be that evil. But she was, and after she had done it I saw her face as she regarded Mary. She thought Mary knew as well and she would have got rid of her too, and so, yes, I called Hettie back because Hettie loved them and she would have stopped Mrs Argylle. Then you came and spoiled it.' Her eyes narrowed. 'I knew that you would not leave well alone.'

'I have learned something of what happened to you as a child, Greer,' I said.

Looking down, she picked at a thread on her shawl.

'I know about your mother and what happened to her, how they treated her. I know that the Gillies sisters made a pet of you but were unkind too.'

She sat so still, so heavy, but her sadness bruised the air.

'I am sorry for how that must have been. I know what it is to lose a mother, Greer.'

She watched me sullenly.

'Life was not easy for you. Will you tell me what happened all that time ago?'

She opened then closed her mouth, but her need to confess was too strong.

'I hurt Miss Gillies and Evangeline. I'm the reason they fell out and the reason she got the burn.'

Her grey eyes barely moved. 'There was a house of dolls.' Her look was wistful. 'I had never loved a thing so much as that. It was Iskar, and in my games the dolls were all those who lived here. I imagined that I were living here too, with all the warmth and riches. But at night my mother wept.' She paused. 'She was so unhappy. Then my mother killed herself, here at Iskar.' The room seemed to breathe; the candle shivered on its wick. I could not take my gaze from her.

'And I took the dolls and when I imagined they were Evangeline and Miss Gillies, I put all my rage into those tiny bodies.' She stopped as though shocked by her own words. 'And that rage somehow went into them, and things went bad – bad between them when they had been friends before.' She paused. 'But they were kinder to me than ever, all of them, all of the Gillieses, and they gave me a home and trained me up to be Miss Gillies' own maid. They taught me my letters, and later I regretted it.'

'You used the dolls again when I was here?'

She flushed. 'It helped to bring them here.'

'And the ones of Mary and me?'

'Not to harm you. To do the opposite – I left you

both safe beneath the bed in the doll's house – only you took them.'

The boards creaked as she shifted her weight. 'I did not like that I had had the power over the dolls. It was wrong, and that is the confession I took to the minister. I told him everything. Nearly everything.'

'What didn't you confess?'

'The widows' whistle.'

And I felt its weight in my pocket and recalled the way it had moulded to my desire and shuddered.

'What did Argylle say?'

She shrugged. 'He was kind. He told me to make amends as best I could and to pray for forgiveness. He told me to keep my silence.' She looked up. 'Perhaps that was wrong, but I believe that he wanted to protect me.'

'How did you come to have a widows' whistle?'

'My grandmother had gifts.' Her eyes dared me to challenge her. 'She told me many things and once she showed me what she kept buried in a box beneath some rocks. She told me what it was for and showed me how to make one. I loved my mother.' For the first time, there was a crack in her voice. 'I found the box and took the widows' whistle. I called my mother back.' Her pale face whitened further. 'But a ghost is not a person. They come back as only shadows and they are not happy. Not happy to be called from rest, and so I had her return.' She stopped talking; the memories pressed into the air. 'That was the secret that I dared not tell him. And that secret must stay so.' She looked at me then, her expression opaque.

'I took the whistle from the doll's house. Did you make another?'

'There was bone and skin that I hid in William's room. It was not hard to fashion a new one.'

I recalled that moment in the graveyard just before I had been pushed, the sound of the footstep in the wet grasses, and knew it had been Bridget who caused me to fall. I remembered the lullaby coming through the mist. Hettie had come not to harm me but to save me.

In my mind's eye I saw the photograph of them, the one Miss Gillies had sent; the way Hettie's arm draped across the children's shoulders. It had not been with possession, I realized now, but love.

'And William? Did you call him too?'

She shook her head. 'Perhaps they were bound together. I do not know. There is magic there I do not understand.'

'I only ever wanted to help Mary,' I said.

'We did not need you here. I would have looked after her. I would have kept her safe.'

'Did you take Mary to the Fiaclach the night before we both fell ill? Was that you?'

'I had to continue to do Mrs Argylle's bidding, but no, I would not have done that and she did not ask me. If someone took Mary there, it was her. It would not be the first time that she had gained access. The house is never locked.'

'But how would she have got Mary away without Mary knowing her?'

She flushed. 'Mrs Argylle gave me chocolate that I was to leave for Mary – those nights, she would have nightmares and sometimes fell so deeply into sleep that Mrs Argylle could have carried her anywhere without waking.'

I saw again Miss Gillies' disgust and Mary trapped in the terror of her dreams. The nightmares had been Bridget's doing too and I was repulsed to the core.

'I believe Mrs Argylle wanted Miss Gillies to send her away,' she continued. 'Although she killed William, I believe it did not sit well with her. And if you had not interfered, it would have happened sooner and Mary would have been safe.'

'Safe? Safe in an asylum? Condemned to a diagnosis of insanity? Miles from friend or family?'

She regarded me stonily.

'You could have made a friend of me, Greer. I would have helped.'

'Make a friend of you? I saw the way you looked at me, just like the rest with all your bookish ways. What would you have said if I told you about the widows' whistle or murder?'

Shame seeped through me. She was right. The glass rattled in its frame as the wind rushed outside. She leaned forward and I caught some of her rage. 'But you sent them away. It was you, wasn't it? I saw your fear when my Hettie returned, and then I saw your triumph when she had gone.' There was a rent in her voice and I glanced away.

'And now?' I said.

She stood and began to comb her hair. She moved slowly and without haste, as she might at the end of any day, her feet solid on the boards, her fingers measured.

'I don't intend to tell Miss Gillies of tonight's events, or your part,' I said. 'What the island will surmise if Bridget's body washes up, I do not know.'

'It will not.' And there was a certainty to her look. that I did not question I put my hand into my pocket and drew out the whistle. It lay light on my palm, innocuous. It was hard to imagine how much could be achieved by the mere blowing of it.

'This belongs with you, Greer. I believe you would not use it ill.'

Her eyes widened and, slowly, she nodded.

I waited, but she began taking out her night things. She did not want me there.

'Thank you, Greer,' I said and went to the door.

But as I pushed it open, I hesitated and retraced my steps. She faced me then and stilled. I went up to her and put a hand on her arm. It was warm and smooth and I saw the darker flecks in her irises. 'You're too good for Paterson. We could be friends now, Greer, couldn't we?'

She regarded me and some of her hostility came to rest in her eyes. 'Goodnight, miss,' she said.

And so I left her, closing the door behind me, and found my way back to the stale breath of the kitchen and to the hall where the fires were now cold, and I climbed the stairs. At the top, I steadied myself on the newel post. Above me, the lamps glowed, half asleep, their yellow light spilling dream-like on the corridor. I felt my skin as if it were new and marvelled at how it had held me through all that had been so dreadful but also all that had brought delight. I felt the impermanency of life and the steadfastness of love. I remembered those minutes in the water and the moment that I had heard Clara. I paused and re-imagined her, as she had been, my sister, my friend, my beloved.

The house was quiet now, returned to timbers and mossy tiles, to chimneys blackened by smoke and dust and faded carpets. A house where Mrs Lenister hummed cheerily in the great kitchen and Miss Gillies reigned in quiet but dignified disappointment. A house with generations of secrets and all the heart's joys and sorrows held only by crumbling walls. I passed through the corridor as if in a dream and to my own room.

Mary was standing at the window waiting for me and I went and stood beside her and gazed out to where the moon cast the shadow of the church across the valley.

'They are there,' she whispered.

And finally I saw them, as clear as a reflection in glass: Hettie with her curled hair and pale cheeks, and a boy whose face I knew as well as I ever could without having seen it in the flesh.

Mary leaned into me. I could not draw my eyes from their figures. Occasionally a breeze caught Hettie's dress, but there was a stillness about them so profound that it did not obey the normal laws of physics.

'Hettie and William,' Mary said, laying her cheek against me, and I slipped my arm around her.

'Hettie and William,' I repeated, but as I watched, their figures grew thinner, becoming more like the bending grass and wind, more like the shape of the sea behind, more like the tussocked ground, until finally, the ridge stood empty, but for the blazing wonder of the moon.

Acknowledgements

So many people have helped bring *The Whistling* from its first inception to the final product.

Firstly, and as always, I want to thank my treasured friend, wonderful writer and astute beta reader, Anita Sloan, for all her support and for the help that she has given at every stage of my writing journey and beyond. I cannot imagine writing without the outlet of our friendship and encouragement.

Many thanks to the fabulous Giles Milburn of the Madeleine Milburn Agency who loved and believed in *The Whistling* from the start and who secured a deal with my dream publisher.

I am indebted to the talent of the lovely Clio Cornish for her enthusiasm and vision and for taking *The Whistling* and turning it into something much better. Thank you also Clio for being so charming about my regular decimation of the track changes.

Thank you so much to Jen Breslin, Ella Watkins, Maddy Woodfield, Deidre O'Connell and the sales department, and the whole team at Michael Joseph for all their fabulous work and enthusiasm.

To Sally Felton, I thank you for the gorgeous proof cover and video that is beautiful and creepy. And to Jessica Hart and Alex Murray for the stunning final cover. I was speechless when I saw them.

Thank you to my copy-editor, Sarah Bance, for her attention to detail and perfect suggestions, and to Nick Lowndes, Jill Cole and Eugenie Woodhouse on the proofreading team.

Writing is a solitary pursuit and finding fellow writers along the way has made so much difference, so I send out massive love to all my writing friends for their generosity and support. A very special shout out to the wonderful, funny, talented and caring people who make up the @virtwriting on Twitter: a truly fabulous group of dedicated writers and friends. My journey would have been so much poorer without you.

Thank you also to the many non-writing friends who have read, offered advice and supported me along the way.

I want to thank my great-grandmother Violet Evangeline O' Donoghue. You know why.

Thanks always to Hugh and my sons who never question my need to write and for all their love and kindness, and especially the cups of tea – more tea?

And last, but very much not least, I am grateful every single day for my precious sisters, Louise and Julia Kelly.

be possible, therefore, to predict CAM-cycling in plants with a knowledge of their deuterium and ^{13}C composition.

The Peperomia species reported in this study show variations of the CAM metabolism. Peperomia scandens, which is a very succulent epiphytic species native to Mexico, shows attributes similar to CAM. This species frequently shows stomatal closure and little or no gas exchange during the day. At night, there is stomatal opening and the majority of the CO_2 is fixed. Concomitant with a CAM-type gas exchange is a massive diurnal fluctuation of organic acids. These plants tend to have a ^{13}C isotope composition more positive than typical C_3 plants but not as positive as C_4 or CAM plants. On occasion in the laboratory, we have measured gas exchange in P. scandens which is more similar to CAM-cycling than CAM. Thus, we suspect that P. scandens may show CAM or CAM-cycling depending on unknown factors not yet measured, such as age of leaves and/or environment.

Peperomia orba has not shown any CAM or CAM-idling carbon metabolism under our experimental conditions. Gas exchange was similar to C_3, and organic acids tended to be low. ^{13}C isotope composition of P. orba is similar to C_3 plants; however, the deuterium composition (unpublished) is heavy comparable to other Peperomia spp. We do not know if P. orba can show the CAM or CAM-idling phenomenon. The green succulent tissue of the spongy parenchyma of P. orba is not as pronounced as P. scandens and P. camptotricha.

Peperomia camptotricha illustrates CAM-cycling as we understand it. There is C_3-type gas exchange with, on occasion, some CO_2 uptake at night, yet a substantial diurnal fluctuation of organic acids. ^{13}C isotope composition is comparable to C_3 plants or perhaps slightly heavier. In these plants when they are performing CAM-cycling, the organic acid metabolism shows most attributes of CAM. Malic acid fluctuates diurnally, similar to total titratable acidity, and citric acid shows little evidence of fluctuation, a typical CAM response. The major apparent difference between CAM and CAM-cycling appears to be that the exogenous CO_2 uptake in CAM-cycling shows the usual C_3 pattern. This analysis, however, is complicated by the shift toward CAM as leaves mature. In P. camptotricha, CAM-cycling is most apparent in young leaves. As leaves mature, the photosynthetic carbon metabolism tends toward a more CAM-like appearance. Plant and leaf age as a factor in the development of CAM is a common phenomenon having been reported previously [9].

When P. camptotricha is severely stressed, stomata close day and night, and there is virtually no CO_2 uptake. Despite this, there is a continual fluctuation of organic acids indicating a shift to the CAM-idling phenomenon. Thus, P. camptotricha shows CAM-idling as well as CAM-cycling. In the experiment shown in Table I, the plants were droughted for 3 weeks and then rewatered. There was full recovery of all activity when rewatered. We visualize that under natural conditions these epiphytes go through frequent periods of drought in which the tissue dries and the plants shift to CAM-idling. When rewatered, they shift back to CAM-cycling. Perhaps the ecological significance of CAM-cycling in these plants is that they are continually poised to withstand drought by rapidly shifting from CAM-cycling to CAM-idling, depending upon water availability.

Our evidence suggests that the CAM-cycling phenomenon is an important and unique metabolic event associated with Peperomia and certain other species such as some Crassulaceae [15], Welwitschia [16], and Pereskia [10]. Undoubtedly, other species show CAM-cycling as well.

Although at this time we know little about CAM-cycling, it does appear that it is characterized and can be defined on the basis of C_3 gas exchange and organic acid fluctuation comparable to CAM. Detailed studies of the biochemistry as well as ecophysiology will be necessary for a thorough and complete understanding.

Acknowledgments

We thank Janet Hann for laboratory assistance in PEP carboxylase and organic acid assays, and Alfredo Huerta who developed the technique in our laboratory for HPLC analysis of organic acids. Research was supported in part by NSF grant PCM 82-00366.

REFERENCES

1. M.J. DeNiro, Earth Plant Sci. Lett. 54, 177 (1978).
2. H.B. Johnson, P.G. Rowland, and I.P. Ting, Photosynthetica 14, 409 (1978).
3. M. Kluge and I.P. Ting, Crassulacean Acid Metabolism (Springer-Verlag, New York 1978).
4. L. LaPre, Ph.D. thesis, University of California (Riverside, California 1979).
5. J.C. Lerman in: Environmental and Biological Control of Photosynthesis, R. Marcelle, ed. (W. Junk bv. Publ., The Hague 1975) pp. 323-336.
6. M.G. O'Leary, Annu. Rev. Plant Physiol. 33, 297 (1982).
7. C.B. Osmond, Annu. Rev. Plant Physiol. 29, 379 (1978).
8. C.B. Osmond in: Crassulacean Acid Metabolism, I. P. Ting and M. Gibbs, eds. (Amer. Soc. of Plant Physiol., Rockville, Maryland 1982) pp. 112-127.
9. O. Queiroz and J. Brulfert in: Crassulacean Acid Metabolism, I. P. Ting and M. Gibbs, eds. (Amer. Soc. of Plant Physiol., Rockville, Maryland) pp. 208-230.
10. L. Rayder and I.P. Ting, Plant Physiol. 68, 139 (1981).
11. L. Rayder and I.P. Ting, Photo. Res. 4, 203 (1983).
12. L. Rayder and I.P. Ting, Plant Physiol. 72, 611 (1983).
13. L. Sternberg, M.J. DeNiro, and I.P. Ting, Plant Physiol. 74, 104 (1984).
14. S.R. Szarek, H.B. Johnson, and I.P. Ting, Plant Physiol. 52, 539 (1973).
15. J. Teeri in: Crassulacean Acid Metabolism, I. P. Ting and M. Gibbs, eds. (Amer. Soc. of Plant Physiol., Rockville, Maryland 1982) pp. 244-259.
16. I.P. Ting and J.H. Burk, Plant Sci. Lett. 32, 279 (1983).
17. I.P. Ting and M. Gibbs (eds.), Crassulacean Acid Metabolism (Amer. Soc. of Plant Physiol., Rockville, Maryland 1982) pp. 316.
18. I.P. Ting and L. Rayder in: Crassulacean Acid Metabolism, I. P. Ting and M. Gibbs, eds. (Amer. Soc. of Plant Physiol., Rockville, Maryland 1982) pp. 193-207.
19. D.J. van Willert, E. Brinckmann, B.M. Eller, and D. Scheitler, Photo. Res. 4, 289 (1983).

Characteristics of CO_2 Fixation and Productivity of Corn and Soybeans

A. Lawrence Christy, Dane R. Williamson

Monsanto Agricultural Products Co., 800 N. Lindbergh Blvd., St. Louis, MO 63167, USA.

INTRODUCTION

Grain yield is an integration of a large number of variables including weather, soil, fertility, genotype and physiology. Once the seed is planted, several of these variables are set and yield will primarily be the result of the physiology of the crop interacting with the weather. Photosynthesis has long been assumed to be one of the key physiological processes and the relationship between crop productivity and photosynthesis, has been the subject of recent reviews [7,8]. Obviously, there is some relationship between photosynthesis and yield, however, to date, the exact nature of this relationship has remained elusive. Recently, we [3] reported that yield was strongly dependent on seasonal photosynthesis in soybean in studies with plant density and shade treatment. Further studies with different cultivars of soybean and corn hybrids have revealed a more complex relationship between grain yield and canopy photosynthesis. The following is a summary of these studies, and we will compare and contrast the characteristics of photosynthesis, photosynthate partitioning, and yield in corn and soybeans.

Canopy photosynthesis measurements were made using chambers developed by Peters et al. [9] and modified for use on corn. The change in chamber CO_2 concentration with time as measured by an infrared gas analyzer, was used to compute the photosynthetic rate.

DIURNAL TIME COURSE OF CANOPY PHOTOSYNTHESIS

On most days, the diurnal cycle of light intensity appears to drive the major environmental changes in the crop canopy [3]. Figure 1 is the time course of corn and soybean net canopy photosynthetic rate for a typical day in mid-summer. The canopies of both crops were closed at this point in the growing season with 100% light interception. Some slight cloudiness at 0830 and 1000 reduced photosynthesis for a short period during the morning hours. A large cloud bank came through at 1500, reducing the photosynthetic rate for soybeans and dropping the corn photosynthetic rate below the light compensation point. Corn appears to be more sensitive to the light intensity changes than soybeans.

Light use efficiency, expressed as grams of CO_2 fixed per μEinstein (PAR) of light striking the top of the canopy, for the data in Figure 1 is plotted in Figure 2. Light use efficiency for both crops remains fairly constant with corn enjoying an approximately 2 to 1 edge over soybeans. The period of low light intensity at 1500 significantly lowered the efficiency of the corn canopy while having a less pronounced effect on the soybean canopy. These data illustrate some of the differences between C_3 and C_4 photosynthesis. While corn has a higher light use efficiency, due in part to the lack of photorespiration, its efficiency is more sensitive to changes in light intensity. In addition, soybean canopies light saturate at 1300 to 1500 μEinsteins m^{-2} sec^{-1} [3], while corn canopies do not saturate at full sunlight. Thus, a 25% drop in intensity from full sunlight will not affect soybean photosynthetic rates, but will dramatically reduce corn photosynthetic rates, which accounts for part of the effect on light use efficiency.

Paul W. Ludden and John E. Burris, editors
NITROGEN FIXATION AND CO_2 METABOLISM

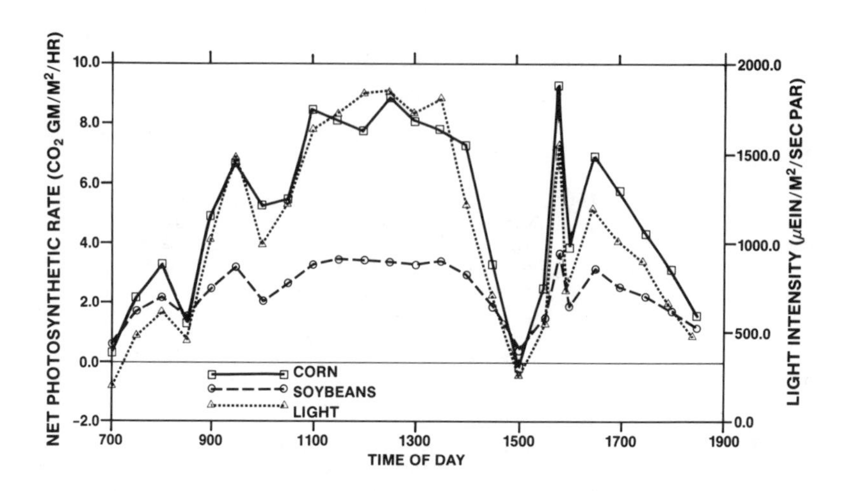

Fig. 1. Time course of light intensity and canopy net photosynthetic rate for corn and soybeans.

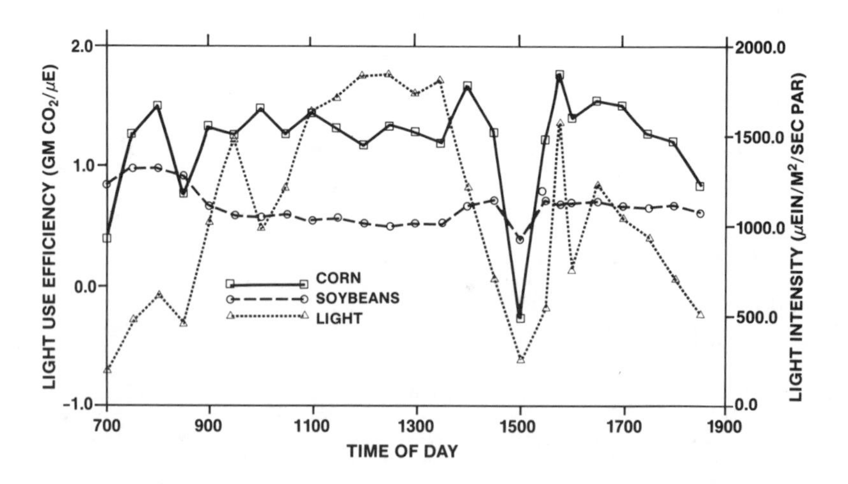

Fig. 2. Time course of light use efficiency for a corn and soybean canopy, computed from the data in Figure 1.

Investigators have reported a decrease or "slump" in the photosynthetic rate at mid-day or early afternoon for soybeans. Over the 10 years we have been studying photosynthesis in the field, we have not observed a mid-day slump in photosynthesis. This may be due in part to the fact that we irrigate our crops. However, even at air temperatures in excess of 40°C, we have not observed a significant decrease in canopy photosynthetic rates.

SEASONAL TIME COURSE OF CANOPY PHOTOSYNTHESIS

A typical seasonal time course of peak canopy photosynthetic rates for corn (PN3780A) and soybean (cv Williams) is shown in Figure 3. These represent peak rates because the data were taken within 2 hours of solar noon and at a light intensity greater than 1500 μEinsteins m^{-2} sec^{-1} on each day. The photosynthetic rate of both crops peak at 35 to 40 days after planting and just prior to the onset of flowering. The photosynthetic advantage of corn over soybean is readily apparent from these data.

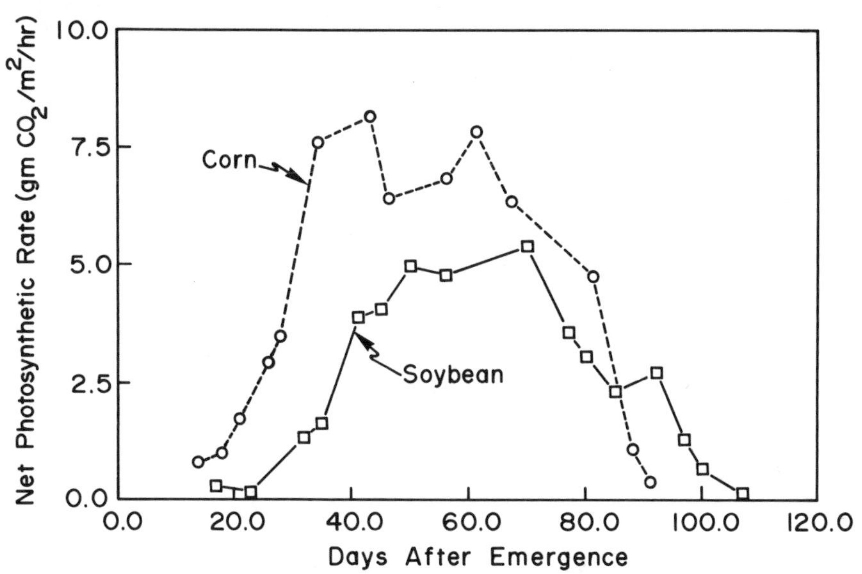

Fig. 3. Seasonal time course of canopy photosynthesis for corn and soybeans.

Estimating Seasonal Photosynthesis

Seasonal photosynthesis can be estimated by integrating the area under the curves in Figure 3. This is only an estimate of seasonal photosynthesis because the data points on this line represent photosynthesis measurements taken every 7 to 10 days during the growing season. This may be an overestimate of seasonal net photosynthesis, because measurements are taken above 75% full sunlight. Fluctuating light intensity under clouds may lead to erroneous measurements of photosynthesis. On days with broken cloud cover, plots measured under full sunlight would be difficult to compare to plots measured under a cloud. In addition, there is some indication that total net photosynthesis on a day with broken cloud cover is the same as the total on a clear day. Overall, we feel our estimate of seasonal photosynthesis is fairly accurate for comparison within a growing season and even between growing seasons.

EFFECT OF GENOTYPE ON PHOTOSYNTHESIS AND YIELD

Soybeans

Varying genotype of the crop has been one method used to study the relationship between photosynthesis and yield [1,11] but has resulted in only limited success. It appears that the strong relationship between seasonal canopy photosynthesis and yield reported earlier [3] does not hold when comparing cultivars of soybean (Fig. 4). The efficiency of converting photosynthate into yield (photosynthetic conversion efficiency or PCE) may be more important in determining yield than we previously assumed [3] (Fig. 5). PCE is computed by dividing grain yield by estimated seasonal photosynthesis [3]. Since the PCE differed for the six cultivars of soybean, there was no apparent relationship between photosynthesis and yield. This may explain why studies using different cultivars have failed to establish a cause and effect relationship between photosynthesis and yield.

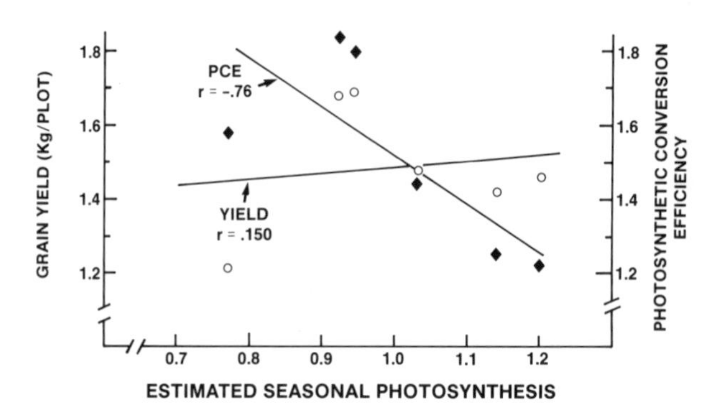

Fig. 4. Yield and PCE as a function of seasonal photosynthesis for six cultivars of soybean.

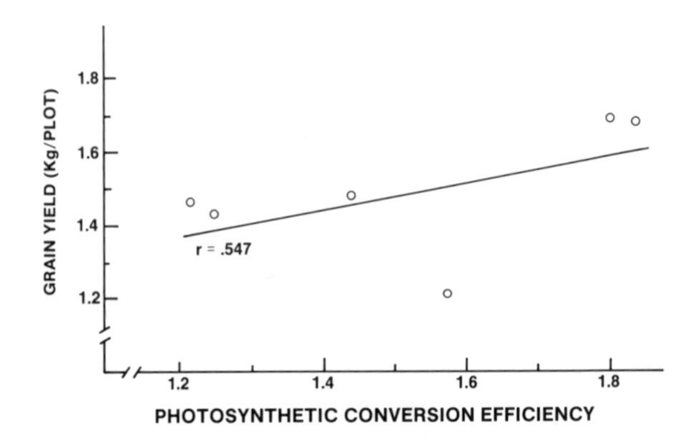

Fig. 5. Grain yield and PCE for six cultivars of soybeans.

The PCE value is both a measure of photosynthate partitioning and an indicator of the efficiency of the crop to convert photosynthesis into yield. It is similar to a harvest index, but includes a better estimate of total dry matter production and not just dry matter at the end of the season. It represents a combination of physiological and biochemical processes that contribute to yield.

Corn

Over the past five decades, corn yields have been increasing at an average rate of 1 bushel per year. Duvick [6] has found that approximately one half of this yield increase can be attributed to plant breeding and one half to changing culture practices. We have just completed a two-year study of 10 corn hybrids, provided to us by Dr. Duvick of Pioneer Hi-Bred International. Two each of these 10 hybrids were introduced in one of the last five decades. The results of both years were very similar and the results from 1982 are shown in Figure 6.

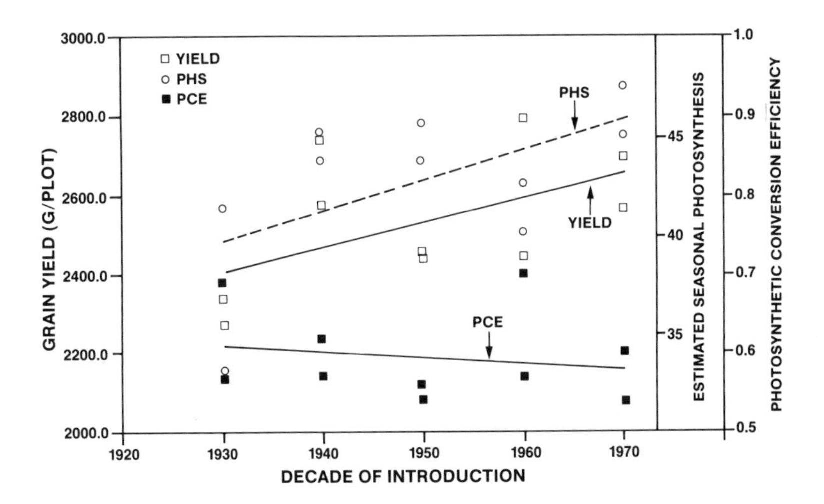

Fig. 6. Grain yield, estimated seasonal photosynthesis, and PCE for 10 hybrids of corn introduced during the last 5 decades.

Although there is an increase in both seasonal photosynthesis and yield over five decades, this increase is not significant. The slight negative slope in the PCE line is due to a larger increase in the seasonal photosynthetic rates than in yield over the five decades, but it also is not statistically significant. We did observe an increase in yield from 1930 to 1940 and an increase in seasonal photosynthesis from 1930 to 1940, and 1960 to 1970. The timing of the sharp yield increase agrees with a recent study of old hybrids by Tapper [10] and indicates a possible relationship between photosynthesis and yield.

The reason we did not observe as large a yield gain as that reported by Duvick [6] may be a combination of location and plot size. Duvick's results indicated that the main effect breeders have had on corn was to make it more tolerant of high population. Perhaps our location further south than where these hybrids are normally grown and in combination with a small plot size negated some of this effect. It is interesting to speculate, however, that the apparent correlation between the increase in grain yield and photosynthesis suggests that corn breeders may have been selecting, at least to some extent, for higher photosynthesis over the last 50 years.

ENVIRONMENTAL EFFECTS

The weather is the single most important factor in crop productivity. One to two week weather patterns can affect yield components, i.e., seed number or seed size, depending on the time during the growing season when these patterns occur [5]. Season-long weather patterns, for example, a hot and dry versus a cool and wet summer, are more influential on the final yield and seasonal photosynthesis. This is illustrated by the data in Table 1 for the corn hybrid PN3780A grown with irrigation for four years in the same field at St. Charles, Missouri. One of the difficulties in relating yield to the weather, is choosing the weather variable that best illustrates the weather pattern. Since our plots were irrigated we choose to use temperature as indicated by the number of days over 30°C (Table 1). Relatively hot dry years occurred in 1980 and 1983, and 1981 and 1982 can be best characterized as cool with adequate rain fall.

TABLE 1. Comparison of canopy photosynthesis and yield components of field grown corn (PN-3780A) over four years.

	1980	1981	1982	1983
Yield (Bu/Acre)	115.1	151.4	190.0	106.2
Seed # (#/M^2)	3171	3043	3622	2481
Seed Size (gm/100)	19.0	26.1	27.4	22.3
Estimated Maximum Potential Seasonal Photosynthesis	36.3	28.9	45.7	31.2
Photosynthetic Converstion Efficiency (PCE)	1.66	2.76	2.17	1.78
Number of Days Above 32°C	62	18	20	63

Yield and PCE appear to correspond to the weather pattern (as indicated by the number of days over 30°C) with the highest yield in 1981 and 1982. The correlation coefficient for a regression of yield and PCE on days over 30°C is -0.89 and -0.88, respectively, and is significant at the 0.11 level. However, the seasonal photosynthesis values do not appear at all related ($r = -0.25$) to the temperature patterns.

While space does not permit an extensive discussion of grain yield and weather interaction, these data do illustrate the significant influence the weather exerts over crop yield. It appears that the main influence of the weather on corn yield may be through PCE. The lower yields in the hotter years of 1980 and 1983 may be due in part to higher dark respiration rates "burning off" photosynthate on warm nights. This pronounced effect of the weather on PCE adds another variable to experiments attempting to establish a relationship between grain yield and photosynthesis.

EFFECT OF PLANT DENSITY

Varying plant density or population is another treatment used to vary yield and canopy photosynthesis [3,4]. We have reported [3] that yield appears strongly dependent on seasonal photosynthesis in soybean with varying population (Fig. 7). A similar experiment was repeated using the corn hybrid PN 3780A in 1980 and 1981. Since canopy seasonal photosynthesis remained relatively constant across the 7 populations while grain yield varied, the PCE had to vary (Fig. 7). Some may object to our plotting grain yield versus PCE because grain yield is used to compute PCE. We choose to plot the data instead of presenting it in a table to better illustrate the relationships between the parameters.

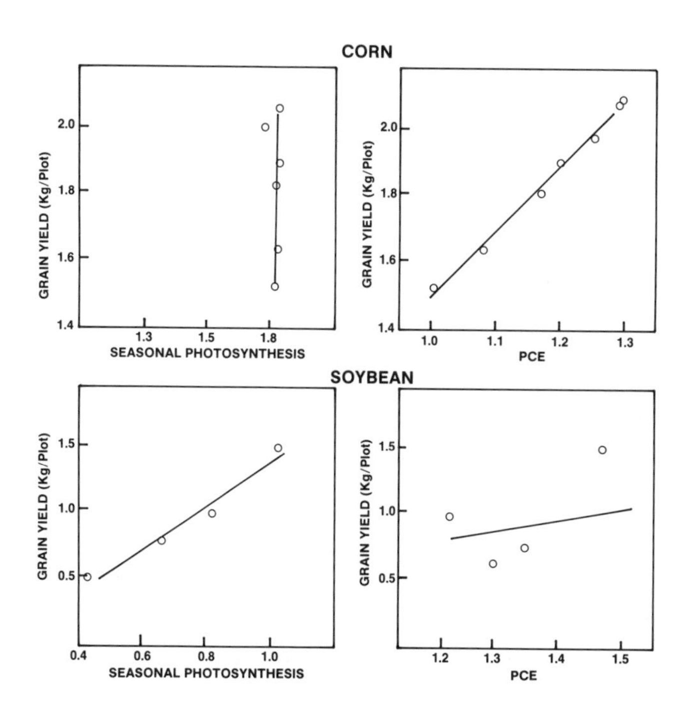

Fig. 7. Grain yield versus seasonal photosynthesis and PCE for different plant populations of corn (PN 3780A) and soybean (cv. Williams).

At first, these data lead us to conclude that corn yield was not related to canopy seasonal photosynthesis and PCE appeared to be the dominate factor controlling yield. However, more recent data from shading [5] and genotype studies (Fig. 5) suggest that photosynthesis may also be limiting yield under certain circumstances. A plot of grain yield on seasonal photosynthesis of the data in Table 1 also tends to support this conclusion (Fig. 8).

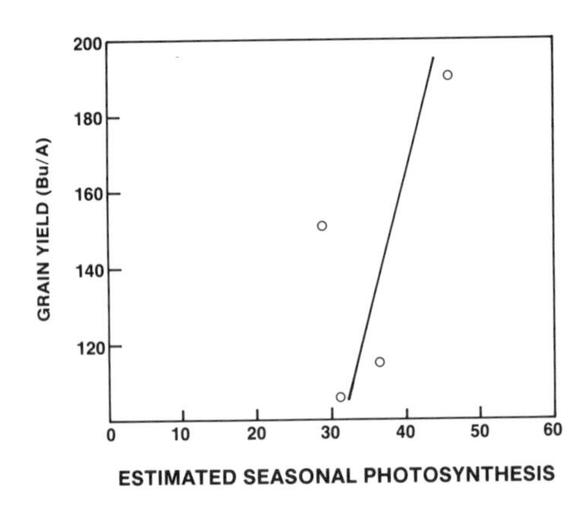

Fig. 8. Grain yield as a function of seasonal photosynthesis for four years (data from Table 1).

While corn yields are not strongly linked to photosynthesis, it plays some role in determining yield. A higher net photosynthetic rate in corn and more stored reserves in the stalk, may help to minimize short-term limitations in assimilate supply. With this reserve source capability, PCE becomes a more influential factor in determining yield in corn.

DISCUSSION AND CONCLUSIONS

Grain yield in corn and soybeans is coupled to seasonal photosynthesis through many physiological processes. These processes can be lumped together as the photosynthetic conversion efficiency (PCE) which is similar to a harvest index and is an indication of overall photosynthate partitioning at the crop level. Lack of data correlating photosynthesis and yield appears to be due in part to our overlooking PCE as a link between photosynthesis and yield. Although harvest index has been important to plant breeders for a number of years, PCE which is a similar variable, was not used to couple yield to photosynthesis until 1982 [3]. The number of variables affecting PCE and differences in the sensitive of PCE in different crops may add to the conflicting ideas concerning photosynthesis and yield.

Corn and soybean appear to "operate" somewhat differently with respect to regulation of yield. In soybean, yield is fairly dependent on seasonal photosynthesis and PCE plays a lessor role. This may indicate that photosynthesis is more limiting in C_3 crops. PCE appears to differ markedly with genotype, but appears less sensitive to population and the environment in soybean. In corn, the relationship between grain yield and photosynthesis appears more difficult to establish or is only loosely coupled. Photosynthesis may not be limiting under "normal" conditions or during certain growth stages. For example, two weeks of cloudy weather during vegetative growth may not affect yield, while a similar period at tasseling may significantly reduce yield.

In corn the PCE appears more variable and is a more influential factor in determining yield. PCE is less dependent on genotype and more sensitive to environment, population, and other external factors than in soybean. Short term changes in photosynthesis due to weather or stress, may be dampened out by PCE in corn. Overall, corn yields appear more dependent on both variables, PCE and photosynthesis, as compared to soybean where yield is more tightly linked with seasonal photosynthesis.

The number of variables and physiologial processes that occur between photosynthesis and yield may account for the lack of data relating photosynthesis and yield. Any treatment used to vary photosynthesis or yield that also affect PCE will lead to confusing results. When you consider the complexity and number of variables involved in this relationship, the lack of data supporting a cause and effect relationship between photosynthesis and yield is understandable. Yield is dependent on photosynthesis, but the strength of this dependence is regulated by many physiological and environmental factors.

REFERENCES

1. R.P. Ariyanayagan, Some Genetic and Physiological factures of Photosynthesis and its Relationship to Yield of Corn (*Zea mays* L.). Dissertation, Cornell University. (1974).
2. R.M. Castleberry, C.W. Crum and C.F. Kroll, Crop Sci. 24, 33-36 (1984).
3. A.L. Christy and C.A. Porter in: Photosynthesis; Vol. II, Development, Carbon Metabolism, and Plant Productivity, Govindjee, ed. (Academic Press, New York 1982) pp. 499-511.
4. A.L. Christy, M.E. Westgate and C.A. Porter. Manuscript in preparation.
5. A.L. Christy and D.R. Williamson, Amer. Soc. Agro. Ann. Meeting (1984).
6. D.N. Duvick, Maydica 22, 187-196 (1977).
7. R.M. Gifford and L.T. Evans, Ann. Rev. Plant Physiol. 32, 485-509 (1981).
8. N.E. Good and D.H., Bell in: The Biology of Crop Productivity, ed. P.E. Carlson (Academic Press, New York 1980) pp. 3-51.
9. D.B. Peters, B.F. Clough, R.A. Graves, and G.R. Stahl, Agron. Jour. 66, 460-462 (1974).
10. D.C. Tapper, Changes in Physiological Traits Associated with Grain Yield Improvement in Single-Cross Maize Hybrids from 1930 to 1970. Dissertation, Iowa State University (1983).
11. R. Wells, L.L. Schulze, D.A. Ashley, H.R. Boerma, and R.H. Brown, Crop Sci. 22, 886-890 (1982).

Dicarboxylate Transport into Isolated Pea Chloroplasts

M. Werner-Washburne, K. Keegstra

Department of Botany, University of Wisconsin—Madison, Madison, WI 53706, USA.

INTRODUCTION

The proper functioning of a plant cell requires substantial fluxes of various metabolites between the cytoplasm and the stromal space of chloroplasts. For some metabolites, such as nitrite and bicarbonate, this transport has been demonstrated to occur by diffusion of the uncharged form of the molecule across the membrane (6). For most metabolites, however, saturation of transport by the substrate and inhibition of transport using various reagents has provided evidence that transport is carrier-mediated.

One of the most active of the chloroplast transport systems is the dicarboxylate transporter. An early model of dicarboxylate transport envisioned a single carrier with broad specificity. According to this model, the dicarboxylate carrier mediates the counter exchange of the dicarboxylates malate, oxaloacetate, 2-oxoglutarate, succinate and fumarate as well as the amino acids aspartate, glutamate and possibly glutamine and aspargine.

More recently, work from several different laboratories has led to the conclusion that the simple model of a single dicarboxylate carrier is inadequate. The remainder of this manuscript discusses some of the evidence that supports the argument for multiple dicarboxylate carriers.

THE DICARBOXYLATE CARRIER

Transport of dicarboxylates and dicarboxylic amino acids into spinach chloroplasts was first described by Heldt and Rapley (7) using the silicone oil centrifugation technique. Later, Lehner and Heldt (9), also using silicone oil centrifugation, characterized the dicarboxylate carrier in some detail with respect to competitive interactions between dicarboxylates and dicarboxylic amino acids. Based on this work and preliminary work, also with spinach chloroplasts, by Gimmler _et al_. (5) the dicarboxylate carrier has been assumed to transport the wide range of dicarboxylates listed above. Because of differences between Km and Ki values for several dicarboxylates, including aspartate, Lehner and Heldt suggested (9) that multiple dicarboxylate carriers may exist. Although this hypothesis is attractive, especially in light of the broad substrate specificity shown by the carrier, no direct evidence for the hypothesis was presented.

Recently Woo _et al_. (17) reexamined the relationship between the transport of 2-oxoglutarate and succinate into spinach chloroplasts. Differences between the Km for succinate transport and the Ki of succinate inhibition of

Published 1985 by Elsevier Science Publishing Co., Inc.
Paul W. Ludden and John E. Burris, editors
NITROGEN FIXATION AND CO_2 METABOLISM

2-oxoglutarate transport provide preliminary evidence that succinate may interact with more than one binding site on the same carrier or with multiple carriers. In addition, the competitive inhibition of succinate transport and noncompetitive inhibition of 2-oxogluarate transport by malate, and the noncompetitive inhibition of succinate transport and uncompetitive inhibition of 2-oxoglutarate transport by ammonia further support the hypothesis that succinate and 2-oxoglutarate have different binding sites.

Although the results of Woo _et al_. (17) are not sufficient to prove the existence of multiple dicarboxylate carriers, they do provide preliminary evidence that dicarboxylate transport is more complex than the simple carrier model would predict. To confirm the presence of separate carriers, differences in inhibitor or substrate specificity between the carriers must also be demonstrated.

In 1981, Proudlove and Thurman (12) had reported that 2-oxoglutarate uptake into peas showed typical Michaelis-Menten kinetics. However, in a more recent study of 2-oxoglutarate transport into pea chloroplasts, Proudlove _et al_ (13) found good evidence for multiple 2-oxoglutarate transport systems. Their results demonstrated that 2-oxoglutarate can completely inhibit malate transport into pea chloroplasts, but that malate can inhibit only a portion of 2-oxoglutarate transport. They concluded that at least two translocators, with similar Km values for 2-oxoglutarate but differing in their ability to bind malate, are present in pea chloroplast envelopes.

Somerville and Ogren (14) were also able to address the question of multiple dicarboxylate carriers during their study of conditional lethal photorespiratory mutants of _Arabidopsis_. One mutant, CS156, was found to carry a recessive nuclear mutation which causes yellowing and a loss of vigor after 3-4 days under standard atmospheric conditions. This mutant accumulates high levels of glutamine and ammonia and decreased levels of glycine and serine under photorespiratory conditions.

Isolated chloroplasts from CS156 showed a marked decrease in malate, 2-oxoglutarate, glutamate, and asparatate transport but no decrease in glutamine or phosphate transport. In order to measure saturable dicarboxylate transport, mutant chloroplasts needed to be preloaded with the substrate to be assayed. This suggests that the saturable dicarboxylate transport system in mutant chloroplasts mediates counter-transport. Furthermore, in the absence of preloading, aspartate uptake into mutant chloroplasts is slow and apparently unsaturable, indicating the presence of possibly a third, low affinity, aspartate transport mechanism.

The mutant and wild type differ greatly in their rates of dicarboxylate transport. However, rearrangement of Somerville and Ogren's data into reciprocal plots suggests the apparent Km values for aspartate transport into both mutant and wild type chloroplasts are similar. Because of this, one would expect, if the mutant transport does represent a second, high affinity carrier, that these carriers would not be kinetically distinguishable in the wild type. Further characterization of transport in the mutant and the wild type, especially with respect to inhibitor sensitivity and substrate specificity, is required to determine whether the observed differences are due to multiple high affinity transport systems.

The high rate of glutamine transport in the *Arabidopsis* dicarboxylate mutant indicates that glutamine is not transported predominantly via the dicarboxylate translocator. This is in contrast to data presented by Gimmler *et al*. (5) and Barber and Thurman (2) which suggested that glutamine and other dicarboxylates are transported via the same carrier. For both results to be correct, glutamate, aspartate, and malate must be transported into the chloroplast by two carriers, one of which also transports glutamine.

The lethality of the dicarboxylate mutant under conditions which favor high rates of photorespiration suggests that a primary function of the dicarboxylate carrier is to transport metabolites involved in photorespiration, i.e., presumably aspartate, malate, glutamate and 2-oxoglutarate. A secondary function is suggested by the fact that under nonphotorespiratory conditions, the dicarboxylate transport mutant fixes CO_2 at a rate which is 140% that of the wild type.

Despite this recent evidence for the heterogeneity of dicarboxylate transport, the model of a single carrier that translocates a broad range of dicarboxylates has not been significantly modified. In each of the three papers discussed above, the evidence for multiple carriers is based on a single line of experimentation and, in every case but that of Proudlove *et al*. (11), could alternatively be interpreted as a function of a single carrier. The evidence, therefore, supports but is not conclusive proof of the hypothesis that multiple dicarboxylate carriers exist. For this reason, the single carrier model, albeit paradoxical, is still generally accepted. The paradoxical nature of the dicarboxylate translocator model is perhaps best understood by comparing the model for chloroplast dicarboxylate transport with that proposed for an analogous organelle, the mitochondrion.

MITOCHONDRIAL DICARBOXYLATE TRANSLOCATORS

Six of the 12 or 13 known mitochondrial translocators mediate the transport of substrates thought to be carried by the single chloroplast dicarboxylate translocator (Table I). All but one of the carriers involved in mitochondrial dicarboxylate transport have been demonstrated to mediate counterexchange (3,8). The glutamate translocator may exchange glutamate for a hydroxide ion, but this type of transport is indistinguishable from the cotransport of glutamate with a proton. Of the six dicarboxylate carriers, three are involved in glutamate transport. Five of the six carriers mediate electroneutral exchange, and one, the aspartate/glutamate translocator, is electrogenic. Of the 5 electroneutral transport reactions, 3 are proton compensated and, thus, would be expected to be sensitive to a pH gradient across the inner mitochondral membrane and to media or internal pH.

TABLE 1. Mitochondrial translocators involved in transporting metabolites thought to be transported via the dicarboxylate translocator in chloroplasts (3,8)

Translocator	Typical species translocated media⟷matrix	Other species translocated
Dicarboxylate	malate^{2-}⟷phosphate^{2-}	fumarate, succinate oxoloacetate
Tricarboxylate	citrate^{3-}+H^{+}⟷malate^{2-}	isocitrate phosphoenol-pyruvate
Malate/ 2-oxoglutarate	2-oxoglutarate^{2-}⟷malate^{2-}	
Glutamate/ aspartate	glutamate^{-}+H^{+}⟷aspartate^{-}	
Glutamate	glutamate^{-}⟷OH^{-}	
Glutamine	glutamine^{+}⟷glutamate + H^{+}	

As seen in Table I, several mitochondrial carriers translocate the same metabolites and, in this way, are said to be linked. For example, the dicarboxylate translocator is linked to the phosphate translocator (not shown) via phosphate transport. Likewise, the tricarboxylate carrier, which transports citrate and malate, is linked directly to the dicarboxylate translocator and indirectly to the phosphate translocator through the transport of malate. Finally, the glutamine/glutamate and glutamate/aspartate carriers are both linked by glutamate to the glutamate carrier. As might be expected, these linkages have sometimes led to confusion in the identification and characterization of separate carriers. For example, because anions such as malate and succinate were found to accumulate in the mitochondria in response to a pH gradient, they were assumed to be transported with a proton. Later, it was demonstrated that transport of these metabolites was indirectly coupled to the pH gradient by phosphate (8). Because phosphate is counter exchanged with a hydroxide ion via the phosphate translocator (not shown) its accumulation in the mitochondria is sensitive to a pH gradient (8).

It is apparent from this brief discussion that the information available on mitochondrial dicarboxylate transport is comparatively sophisticated. In contrast, studies of chloroplast dicarboxylate transport are just beginning to deal with approaching the question of multiple dicarboxylate carriers.

THE STUDY OF L-ASPARTATE TRANSPORT INTO PEA CHLOROPLASTS

To address the question of multiple dicarboxylate carriers, we chose to study L-aspartate transport into isolated pea chloroplasts. L-Aspartate was selected for several reasons. First, aspartate is thought to be translocated solely by the dicarboxylate carrier. In addition, it has a significant role in both chloroplastic and cellular metabolism. It is required for protein synthesis in chloroplasts, mitochondria, and the cytoplasm and is a precursor for the synthesis of several other important chloroplastic and cytoplasmic compounds including the essential amino acids, lysine, threonine, isoleucine, and methionine (11,16).

Aspartate has also been postulated to function in a shuttle for exporting reducing equivalents from the chloroplast to the cytoplasm during photosynthesis (4) and to the glyoxysome during photorespiration (15). Previously, a malate/oxaloacetate shuttle was postulated to shuttle reducing equivalents, however, a recent study by Giersch (4) suggests that, based on the respective Km and Vmax values and the chloroplastic and cytoplasmic concentrations of both malate and oxaloacetate, this shuttle is not functional *in vivo*. Because of the relatively high cellular concentrations of aspartate, if such a shuttle does exist, it is likely to involve aspartate. In the shuttle of reducing equivalents, aspartate is postulated to be deaminated and reduced to malate in the chloroplast, malate transported to the cytoplasm for subsequent oxidation to oxaloacetate, and oxaloacetate transaminated to aspartate for transport back to the chloroplast. As in the mitochondrial malate/aspartate shuttle, a parallel glutamate/2-oxoglutarate shuttle could function in maintaining a nitrogen balance between the chloroplast and the cytoplasm.

Because of the involvement of aspartate in such a wide variety of metabolic pathways in both the cytoplasm and the chloroplast, it seemed likely that several chloroplast carriers might function in its transport. In addition, previous work by Lehner and Heldt (9) had pointed to the possible existence of multiple aspartate carriers. We chose to study the uptake kinetics and inhibitor sensitivity of L-aspartate transport into isolated pea chloroplasts in the hope that these two transport functions might provide a basis for the identification of multiple carriers.

Our results indicate the presence of at least two aspartate carrier systems distinguishable both kinetically and by their differential sensitivity to inhibitors. The first is a high Km, or low affinity, transport. The second system is a low Km, high affinity, carrier-mediated transport. Preliminary work with nigericin, p-chloromercuribenzenesulfonate (PCMBS), and diethylpyrocarbonate (DEPC) suggest that these inhibitors may be useful for further dissection of high affinity transport.

High and Low Affinity Aspartate Transport

Using the silicone oil centrifugation technique (9) we examined the concentration dependence of L-aspartate uptake into isolated pea chloroplasts. Based on the results of these studies with aspartate concentration from 12.5 μM to 30 mM, we identified two kinetically distinguishable transport components (Fig. 1). The Km of the high affinity component was 30 μM ± 16.1 μM (n=9). The Km of the low affinity component was not determined quantitatively because of the error in measurement of transport from high concentrations of aspartate. As a result of this uncertainty, it was not possible to distinguish on the basis of kinetic plots whether low affinity transport was carrier-mediated or represented simple diffusion.

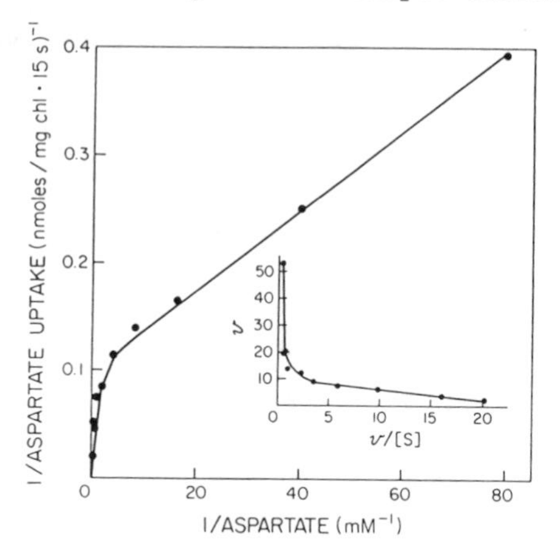

Figure 1. Reciprocal and Eadie-Hofstee plots of L-aspartate uptake into pea chloroplasts (*Pisum sativum* var. Laxton's Progress no. 9). Each point represents the mean of three 15s assays. Chloroplasts were isolated as described in Cline *et al*. (1) and transport measured at 4°C under fluorescent laboratory lights (20 $\mu E/m^2 \cdot s$). Buffer pH was 7.6. Aspartate concentrations used in this experiment ranged from 12.5 μM to 10 mM.

Inhibitor Studies

Several classes of inhibitors were tested for their effects on high affinity aspartate transport. Dipeptides were ineffective as inhibitors whereas the sulfhydryl reactive reagents, PCMBS and mersalyl; the histidine reactive reagent DEPC; and the protonophores, nigericin and carbonyl cyanide M-chlorophenyl hydrazone (CCCP) were effective inhibitors of high affinity transport.

As seen in Table II, of the inhibitors tested, only DEPC inhibited low affinity aspartate transport. The inhibition of low affinity transport by DEPC is currently the only direct evidence that low affinity transport is carrier-mediated. Finally, the differential inhibition of high and low affinity transport by both PCMBS and nigericin provides good evidence that these transport components represent separate carrier systems.

TABLE II. Inhibition of high and low affinity aspartate transport by various reagents. Pea chloroplasts were prepared and assayed as described in Figure 1. Inhibitors were incubated for 20 minutes with chloroplasts prior to assay.

Inhibitor	(concentration)	Inhibition High Affinity Transport	Low Affinity Transport
PCMBS	(up to 1 mM)[a]	+	-
mersalyl	(400 μM)	+	n.d.
DEPC	(1.4 mM)	+	+
nigericin	(0.1-10 μM)	+[b]	-
CCCP	(4 μM)	+	n.d.

[a]all inhibitors were incubated with chloroplasts and transport assayed at pH 7.6 with the exception of DEPC which was incubated at pH 6 and transport assayed at pH 6 or pH 7.6.

[b]inhibition of high affinity transport by nigericin was approximately 70% and was independent of nigericin concentration.

n.d. not determined

Discussion

The results presented here indicate that at least two aspartate transport components are present in pea chloroplasts. These components are distinguishable kinetically and on the basis of their inhibitor sensitivity.

With respect to transport *in vivo*, the low affinity transport may be physiologically important because chloroplast and cytoplasmic concentrations of aspartate are relatively high. For example, aspartate concentrations in pea chloroplasts have been estimated to be 1 mM (Werner-Washburne, unpublished) to 5 mM (10). With an external concentration of 2 mM aspartate, low affinity transport, measured in this study, can represent up to 50% of the total aspartate uptake. Thus, the low affinity transport component could account for a significant proportion of *in vivo* aspartate transport.

The Km for high affinity aspartate transport reported here (30 μM) is lower than the Km for aspartate transport into spinach chloroplasts (720 μM) (9) or into *Arabidopsis* chloroplasts (500 μM) (14). These differences may reflect variations in experimental conditions or between plant species. With respect to experimental differences, the kinetic values obtained in this study were based on assays using concentrations of aspartate from 12.5 μM to 30 mM, as compared with 100 μM to 1.5 mM used by Lehner and Heldt (9). Because the bend in the reciprocal plot typically occurred from 500 μM to 2.5 mM aspartate, data obtained from the concentration range used by Lehner and Heldt (9) might not show a significant bend and a line drawn through these points would give higher estimates for the kinetic parameters. For example, if the uptake rates for aspartate concentrations from 100 μM to 1.5 mM measured in our study were used for calculating kinetic parameters the estimated Km for high affinity transport would be approximately 200 μM. Thus, it is clear that the concentration range chosen for a particular study can have a great effect on the estimated kinetic parameters.

From our work and the work of Woo *et al.* (17), Proudlove *et al.* (13), and Somerville and Ogren (14) it is clear that dicarboxylate transport is much more complicated than predicted by the single carrier model. Our results and those of Somerville and Ogren (14) indicate the presence of high and low affinity aspartate transport in peas and *Arabidopsis*, respectively. Proudlove *et al.* (13) provided convincing evidence for two high affinity 2-oxoglutarate transport systems in peas. The work of Woo *et al.* (17) and Somerville and Ogren (14) is also consistent with the idea of multiple high affinity chloroplast dicarboxylate translocators.

In order to distinguish one high affinity transport system from another, specific inhibitors need to be identified which can selectively eliminate one translocator at a time. The inhibitors identified in this study, i.e. PCMBS, DEPC, and nigericin may well prove useful for further subdividing high affinity transport. For example, the partial, concentration-independent inhibition of aspartate transport by nigericin may indicate the presence of two high affinity transport components, one which translocates aspartate with a proton and another which transports aspartate without proton cotransport.

In the future, as we learn more about chloroplast transport and the chloroplast envelope, it should be possible not only to distinguish between carrier systems, but also to identify carrier proteins and study them with respect to their responses to changes in pH gradients and ionic conditions, their similarity to translocators in other plant cell membranes or analogous organelles, and their involvement in chloroplast biogenesis. Understanding chloroplast transport and the proteins which regulate this transport is only one aspect of understanding the larger question of how the envelope membranes mediate the integration of biochemical events in both the chloroplast and the cytoplasm. Previously we have lacked sophisticated tools for examining chloroplast translocators. However, with the availability of an Arabidopsis transport mutant, the identification of several new classes of inhibitors, and the hope of using specific inhibitors, antibodies, or mutants to identify carrier proteins, we are moving into a new phase of transport studies.

REFERENCES

1. D.J. Barber and D.A. Thurman Plant, Cell, and Environment 1:297-303 (1978).
2. K. Cline, J. Andrews, B. Mersey, E.H. Newcomb, and K. Keegstra Proc. Nat. Acad. Sci. USA 78:3595-3599 (1981).
3. L. Ernster and G. Schatz J. Cell Biol. 91:227s-255s (1981).
4. C. Giersch Arch. Biochem. Biophys. 219:379-387 (1982).
5. H. Gimmler Planta 120:47-61 (1974).
6. U. Heber and H.W. Heldt Ann. Rev. Plant Physiol. 32:139-168 (1981).
7. H.W. Heldt and L. Rapley FEBS Lett. 10:143-148 (1970).
8. K.F. LaNoue and A.C. Schoolwerth Ann. Rev. Biochem. 48:871-922 (1979).
9. K. Lehner and H.W. Heldt Biochim. Biophys. Acta 501:531-544 (1978).
10. W.R. Mills and K.W. Joy Planta 148:75-83 (1980).
11. W.R. Mills, P.J. Lea, and B.J. Miflin Plant Physiol. 65:1166-1172 (1980).
12. M.O. Proudlove and A. Thurman New Phytol. 88:255-264 (1981).
13. M.O. Proudlove, D.A. Thurman, J. Salisbury New Phytol. 96:1-5 (1984).
14. S.C. Somerville and W.L. Ogren Proc. Natl. Acad. Sci. USA 80:1290-1294 (1983).
15. N.E. Tolbert in: Encyclopedia of Plant Physiology, Vol 6., M. Gibbs and E. Latzko, eds. (Springer-Verlag, New York 1979) pp. 338-352.
16. R.M. Wallsgrove, P.J. Lea, and B.J. Miflin Plant Physiol 71:780-784 (1983).
17. K.C. Woo, U.I. Flugge, H.W. Heldt Proc. 6th Internatl. Congr. on Photosynth., Brussels, in press (1984).

Regulation of Photosynthetic Sucrose Formation in Leaves

Steven C. Huber, Douglas C. Doehlert, Phillip S. Kerr, Willy Kalt-Torres

United States Department of Agriculture, Agricultural Research Service, and Departments of Crop Science and Botany, North Carolina State University, Raleigh, NC 27695–7631, USA.

INTRODUCTION

Starch and sucrose are the principal end products of photosynthesis in most higher plants, and as such account for the majority of the carbon assimilated. Sucrose is the primary transport form of reduced carbon and is exported from mature leaves to other plant parts. The concentration of sucrose in leaves represents the balance between the rates of formation and export. In species such as soybean, which do not accumulate a substantial 'storage' pool of sucrose during the day, the rate of assimilate export provides a good estimate of the rate of carbon flux into sucrose. Starch, in contrast, accumulates in leaves during the day as an insoluble end product; in the following dark period, starch reserves can be mobilized to support continued sucrose synthesis and export in the absence of photosynthesis.

Starch is synthesized in the chloroplast, whereas sucrose formation occurs in the cytoplasm of the leaf mesophyll cell [1]. It is thought that carbon skeletons for sucrose formation are exported from the chloroplasts primarily as triose phosphates. Release of triose phosphates, in strict exchange for the uptake of inorganic phosphate (Pi), is catalyzed by the phosphate translocator of the chloroplast envelope [2]. The triose phosphates originate in the chloroplast from concurrent photosynthesis in the light or from degradation of starch in the dark. As four molecules of triose phosphates are converted to one sucrose molecule in the cytosol, four molecules of Pi are liberated. The rate of sucrose formation will therefore largely determine availability of Pi for counterexchange with triose-P. Thus, the rate of sucrose synthesis, relative to the rate of CO_2 assimilation, may determine the degree to which carbon is diverted from starch biosynthesis in the chloroplast.

Partitioning of photosynthetically fixed carbon between starch and sucrose varies among species [3,4], and with a given genotype can be altered by environmental parameters such as photoperiod [5,6]. Thus, carbon partitioning is clearly controlled independently of photosynthetic rate. Some of this overall control may be at the level of the rate of sucrose formation in the cytosol. Two of the cytoplasmic enzymes that catalyze essentially irreversible reactions and appear to be regulated steps are 1) cytoplasmic fructose-1,6-bisphosphatase (FBPase; E.C. No. 3.1.3.11) and 2) sucrose phosphate synthase (SPS; E.C. No. 2.4.14). A common feature of these two enzymes is that their activities in crude leaf extracts are low in comparison to activities of other enzymes and often similar to the rate of carbon flux into sucrose [7-9]. In particular, changes in the activity of SPS in leaves are often associated with alterations in the partitioning of carbon between starch and sucrose. For example, correlated changes in photosynthate partitioning and SPS activity have been observed during leaf expansion [10,11] and aging [12], by changes in plant nutrition [13], and in response to "source-sink" manipulations [14]. We have postulated that regulation of the activity of SPS in leaves (as assayed under optimal conditions in crude leaf extracts) serves as a "coarse" control mechanism that may contribute to the control of sucrose formation when photosynthetic rates are maximal. In these systems, the

Paul W. Ludden and John E. Burris, editors
NITROGEN FIXATION AND CO_2 METABOLISM

activity of cytoplasmic FBPase usually remains constant, although catalytic activity _in situ_ may be modulated by metabolic effectors [8]. In particular, fructose-2,6-bisphosphate (F26BP) is a recently discovered regulatory metabolite [15,16] that has been shown to occur in leaves and has been reported to be a potent inhibitor of cytoplasmic FBPase [17,18]. Preliminary results suggest that the concentration of F26BP in leaves varies and may play a role in the regulation of carbon metabolism [19,20].

SPS activity is also regulated by metabolic effectors which play an important role in modulating enzyme activity _in situ_ when photosynthetic rates are less than maximal (e.g., limiting light). With purified spinach leaf SPS, Pi is an inhibitor [8,21] and glucose-6-P (G6P) is an activator [22,23], and the two effectors are antagonistic.

In this paper, we compare the regulatory properties of SPS isolated from leaves of different species, and present evidence for diurnal rhythms in SPS activity in soybean leaves as one example of the "coarse" control of enzyme activity that may be related to a physiological process, viz. assimilate export.

RESULTS AND DISCUSSION

Metabolic Regulation of SPS

G6P activation of spinach leaf SPS is observed as a two-fold increase in V_{max} and a five-fold reduction in the apparent Km (F6P) [22]. The apparent Km (UDPG) is unaffected by G6P. Hence, G6P activation is observed regardless of substrate concentrations used to assay enzyme activity, but the relative activation increases as the concentration of F6P is decreased [23]. The effect of Pi on G6P activation at a limiting concentration of F6P is shown in Fig. 1. Phosphate antagonized the G6P activation and altered the shape of the response curve for G6P from apparently hyperbolic to slightly sigmoidal. In the absence of Pi, half-maximal activation by G6P occurred at a concentration of about 0.8 mM, and increased to 9.9 mM in the presence of 20 mM Pi. When the data in Fig. 1 were replotted by the Hill equation, a series of parallel lines were obtained with slopes (Hill coefficients) of about 1.6.

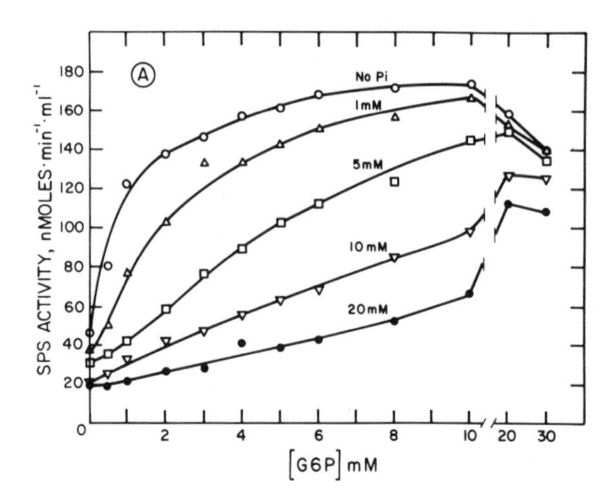

Fig. 1. Effect of Pi on G6P activation of spinach leaf SPS with 8 mM UDPG and 1 mM F6P (limiting). With permission from [23].

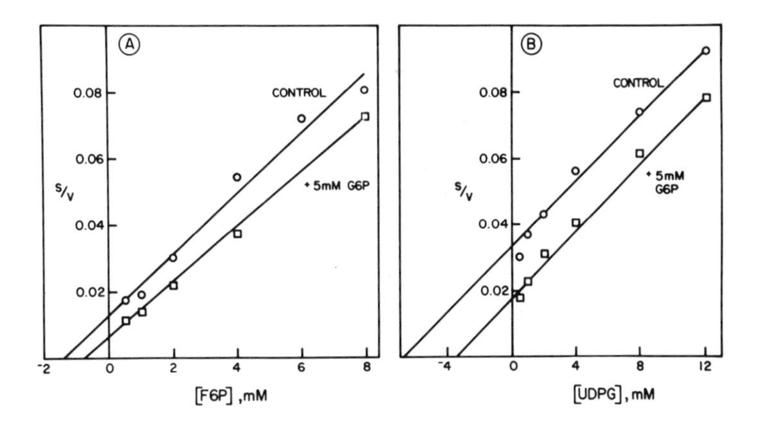

Fig. 2. Woolf plots of saturation curves for A) F6P and B) UDPG for tobacco leaf SPS in the absence and presence of 5 mM G6P. The non-varied substrate was present at a concentration of 8 mM.

In addition to spinach, SPS has been partially purified and partially characterized from several other leaf sources. The response to G6P was diverse. For example, Fig. 2 shows Woolf plots of substrate saturation curves for tobacco leaf SPS in the absence and presence of 5 mM G6P. With tobacco SPS, G6P had little effect on V_{max} but decreased the apparent Km for both substrates. The effect on Km was greatest for UDPG, which was quite high in the absence of G6P (about 6.8 mM) and was reduced to 3.4 mM in the presence of 5 mM G6P. In all cases, substrate saturation curves for both F6P and UDPG were strictly hyperbolic, as previously reported for spinach [8,22,23] and shown in Fig. 2 with the tobacco enzyme.

Table I. Activation by G6P and Sucrose Inhibition of SPS Isolated from Leaves of Different Species and a Nonphotosynthetic Source[a)]

Enzyme Source	Km (F6P) Control	Km (F6P) 5 mM +G6P	Km (UDPG) Control	Km (UDPG) 5 mM +G6P	+ 5 mM G6P	Inhibition by 100 mM Sucrose
	(mM)		(mM)			(%)
Spinach	3.0	0.6	1.9	1.9	2.0	<10
Wheat Germ	1.7	0.9	5.8	6.1	1.6	60
Tobacco	1.4	0.8	6.1	3.7	1.1	20
Barley	0.8	0.7	4.9	3.2	1.0	47
Maize	1.0	1.0	7.0	3.2	1.5	80
Pea	2.1	1.6	3.8	2.8	1.1	53

a) SPS was partially purified from the indicated source using a procedure modified from [23] that involved hydrophobic chromatography, PEG-fractionation, and gel filtration. Except as indicated, enzyme activity was assayed with 8 mM F6P, 12 to 20 mM UDPG, 5 mM $MgCl_2$ and 50 mM Hepes-NaOH (pH 7.5).

Table 1 compares some kinetic and regulatory properties of SPS isolated from different sources. In the absence of the activator G6P, Km (F6P) ranged from 0.6 to 3.0 mM, and the Km (UDPG) varied from 1.8 to 7.0 mM. The effect of G6P on V_{max} and substrate affinities was variable. With spinach, wheat germ, tobacco, and pea, the Km (F6P) was lowered. In tobacco, maize, barley, and pea, the Km (UDPG) was decreased. Thus, with tobacco and pea, the Km for both substrates was affected whereas with the other species, the Km for only one substrate (either F6P or UDPG) was decreased. G6P had a significant effect on V_{max} only in spinach, wheat germ and maize. Hence, with these species the absolute activation caused by G6P was greatest, and relative activation increased as substrate concentration became limited.

Another parameter that varied with the source of the enzyme was inhibition by sucrose. Maize, barley, wheat germ and pea were strongly inhibited by sucrose (>50% inhibition by 100 mM sucrose), whereas spinach and tobacco SPS were markedly less sensitive (Table I). Inhibition of purified wheat germ SPS by sucrose has been documented previously [24], as has the lack of effect on purified spinach SPS [8,21]. However, this is the first report of sucrose inhibition of SPS purified from leaves of different species. Earlier work with crude leaf extracts suggested that sucrose inhibition of SPS varied among species [25]. The variation in properties of SPS from different species suggests a diversity of strategies in the regulation of sucrose formation. The full significance of these differences in properties, however, remains to be determined.

Diurnal Rhythms in SPS Activity

In addition to metabolic fine control of enzyme activity *in situ* by effectors such as G6P, Pi and sucrose, the activity of SPS in leaves is variable and appears to be regulated (coarse control). One example of this 'coarse control' is the diurnal rhythm in SPS activity observed in soybean leaves [26]. SPS activity fluctuates diurnally in leaves of some species (soybean, maize, pea) but apparently not others, such as spinach (manuscript in preparation). In soybean, the rhythm in SPS activity persisted in either extended light or darkness, and thus appears to be controlled by an endogenous clock mechanism [27].

One of our objectives has been to determine the biochemical basis for the observed fluctuation in enzyme activity (i.e., protein modification versus changes in amount of SPS protein). In an attempt to identify changes in activity associated with protein modification, we have monitored several regulatory properties of SPS throughout the diurnal cycle. For example, Fig. 3 shows the diurnal fluctuation in SPS activity in leaves of vegetative soybean plants, assayed in the absence and presence of the inhibitor Pi (15 mM). In general enzyme activity declined continually throughout the day and increased during the night. Pi inhibition was essentially constant at about 45% (Fig. 3 inset). At times of "high" and "low" activity, SPS was partially purified and kinetic constants determined. Even though total activity varied, the apparent Km (F6P) and Km (UDPG) remained constant at 1.9 and 5.5 mM, respectively. Thus, when SPS activity changes diurnally, there are no substantial alterations in substrate affinities or regulatory properties (e.g., Pi inhibition) that could explain the changes in activity or provide positive evidence for protein modification. It is possible, however, that a modification mechanism exists that only affects V_{max}, or that the amount of SPS protein varies diurnally.

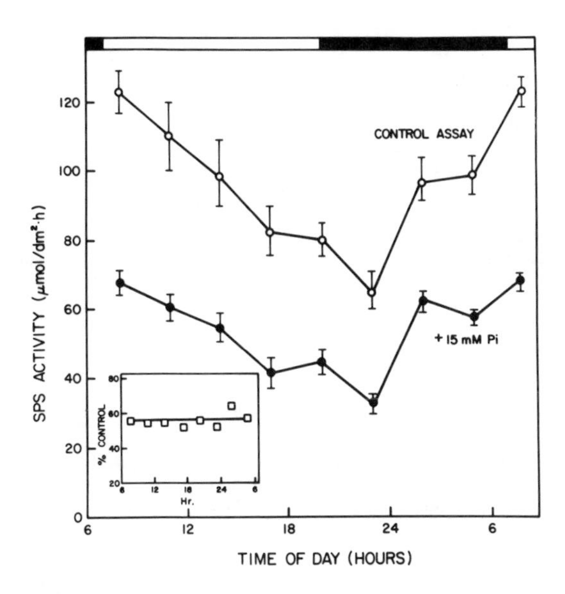

Fig. 3. Diurnal fluctuation in SPS activity in leaves of vegetative soybean plants. Activity in crude leaf extracts (with 8 mM each substrate) was assayed in the absence and presence of 15 mM Pi. Inset: replot of data showing relative inhibition by Pi. The bar at the top of the figure denotes the daily light-dark cycle.

Table II. Variation in photosynthate partitioning and biochemical components of sucrose formation during the photoperiod in soybean leaves.

Time Interval	Mean Rate or Activity During Interval				
	CER[a)]	Export[b)]	Starch[c)]	SPS Activity[d)]	F26BP Concn[e)]
	------(mg $CH_2O/dm^2 \cdot h$)------			($\mu mol/dm^2 \cdot h$)	(nmol/g)
0800 to 1100 h	25.5	16.5	6.0	118	0.13
1100 to 1400 h	28.6	14.5	9.0	106	0.19
1400 to 1700 h	23.5	10.6	11.6	91	0.23
1700 to 2000 h	23.5	11.4	7.5	82	0.26

a) Carbon exchange rate measured during the indicated time interval with a Li-Cor Portable Photosynthesis System; b) Mass carbon export rate calculated by the method of Terry and Mortimer [28]; c) Starch accumulation rate, measured enzymatically [26]; d) Extraction and assay as previously described [14]; e) Measured as in [20].

Another objective of our research is to determine whether the rhythm in SPS activity has some physiological significance. Because SPS activity in leaves is often correlated with the rate of assimilate export [3,4,13, 14], we wanted to determine whether export rate varies diurnally. In vegetative soybean plants, photosynthetic rates remain high and relatively constant throughout the day but photosynthate partitioning changes, as evidenced by the observation that export rate decreases while starch accumulation rate generally increases as the day proceeds (Table II). Thus, it can be inferred that the rate of sucrose formation _in situ_ (which largely determines the rate of assimilate export) varies diurnally. The progressive decrease in the rate of sucrose formation was closely associated with a decrease in SPS activity (related to the diurnal rhythm) and an increase in the concentration of F26BP in leaves (Table II). Thus, the reduction in the rate of sucrose formation in leaves was associated with decreased activities of both cytoplasmic FBPase and SPS. It should be noted that the activity of FBPase in leaves remains constant throughout the diurnal cycle [26,27], but activity _in situ_ may vary as a result of changes in regulatory metabolites as F26BP.

To determine whether the rate of assimilate export displayed a rhythmic pattern over a 24 h period, an experiment was conducted that involved transferring plants at different times of the day and night to an illuminated growth chamber. Assimilate export rate was measured over the 3 hour interval following transfer to the chamber. Figure 4 shows that two distinct peaks in assimilate export were observed when plants were illuminated at different times of the diurnal cycle. Periods of maximum export occurred during the middle of the day and middle of the normal dark cycle. The pronounced fluctuation in export rate was the result of variation in photosynthetic rate (data not shown) and also SPS activity (Fig. 4).

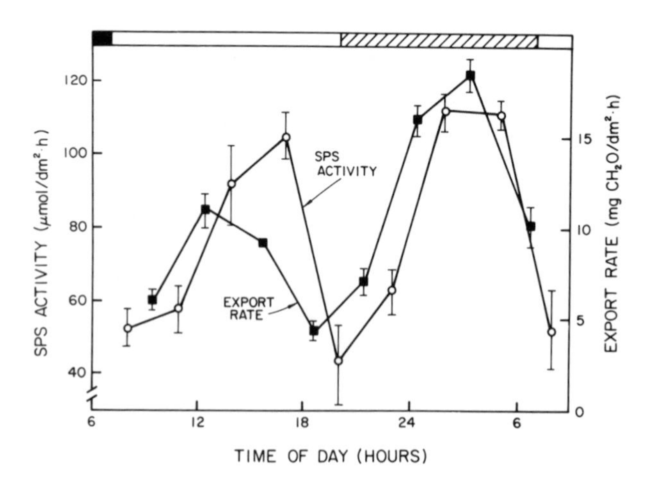

Fig. 4. Diurnal fluctuations in assimilate export rate and the activity of SPS in soybean leaves after transfer of plants from the greenhouse to an illuminated growth chamber (500 $\mu E/m^2 \cdot s$, PAR) at different times of the day and night. Plants were illuminated for a 3 hour interval, during which time export rate was measured [28], and at the end of the interval leaves were harvested for SPS assays.

The diurnal rhythm in SPS activity shown in Fig. 3 for plants maintained in the normal light/dark cycle was obtained with the same population of plants subjected to illumination at different times of the 24 h period (Fig. 4). Thus, the rhythms in SPS activity can be directly compared. The different diurnal rhythms observed suggest that SPS activity in leaves can be affected at certain times of the 24 h period by some environmental factor(s). At the present time, it is not known what factors might be involved, but light intensity, light quality, and relative humidity are possibilities.

Over both types of experiments (Figs. 3 and 4), the rhythms in SPS activity and assimilate export rate were closely aligned, as there was a significant positive correlation between the two parameters (Fig. 5).

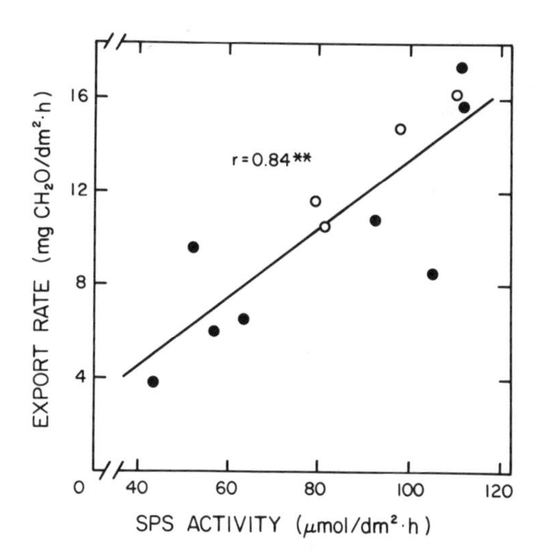

Fig. 5. Relation between assimilate export rate and SPS activity in soybean leaves. Data from Table II and Fig. 4.

Summary

Regulation of the rate of sucrose formation in leaves appears to involve two types of mechanisms, viz. metabolic "fine control" (cytoplasmic FBPase and SPS) and "coarse control" that affects enzyme activity in leaves (SPS). Some of the options for metabolic regulation of the cytoplasmic sucrose formation pathway are summarized in Fig. 6. One major function of the metabolic regulation may be to coordinate the rate of sucrose formation in the cytosol with availability of triose phosphates from the chloroplast (i.e., photosynthetic rate). It is likely that a reduction in CO_2 assimilation rate (e.g., limiting light) would result in an increase in cytosolic Pi as a result of the reduced supply of triose phosphates. Stitt _et al_. [29] have shown that cytosolic P-esters (such as G6P) decrease as CO_2 fixation becomes light or CO_2-limited. An increase in cytosolic (Pi) would activate synthesis of F26BP [30], which then would inhibit cytoplasmic FBPase (Fig. 6). In addition, the reciprocal rise in Pi and decrease in G6P would reduce SPS activity _in situ_, bringing about a

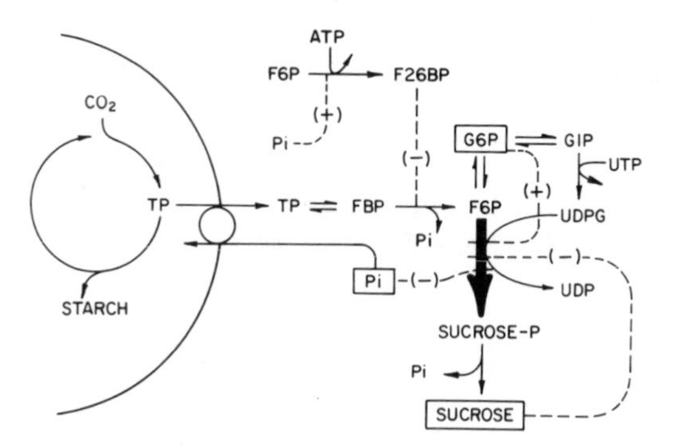

Fig. 6. Pathway of cytoplasmic carbon flow for sucrose synthesis showing regulation of FBPase by F26BP and regulation of SPS (heavy arrow) by G6P, Pi and sucrose. Pi activation of fructose-6-P, 2 kinase from [30]. (+): activation. (-): inhibition. See text for discussion.

concerted reduction in the flux of carbon into sucrose. In some species, SPS is also inhibited by sucrose, which may provide a mechanism to coordinate the rate of sucrose formation with plant demand for assimilates; i.e., accumulation of sucrose in leaves as a result of a decrease in translocation rate would cause feed-back inhibition of SPS and thus reduce the rate of sucrose formation.

When photosynthetic rates are high and sucrose export from leaves is unimpeded, the rate of assimilate export is related to the activity of SPS in leaves. The close correspondence of diurnal rhythms in export rate and SPS activity strongly supports the postulate that coarse control of SPS activity is involved in the control of export rate.

ACKNOWLEDGMENTS

The authors wish to thank Mr. Mark Bickett for excellent technical assistance.

Cooperative investigations of the United States Department of Agriculture, Agricultural Research Service and the North Carolina Agricultural Research Service, Raleigh, NC 27695-7631. Paper number 9329 of the Journal Series of the North Carolina Agricultural Research Service, Raleigh, NC.

Mention of a trademark or proprietary product does not constitute a guarantee or warranty of the product by the U. S. Department of Agriculture or the North Carolina Agricultural Research Service and does not imply its approval to the exclusion of other products that may also be suitable.

REFERENCES

1. S.P. Robinson and D.A. Walker, FEBS Lett. 107, 295-299 (1979).
2. U. Heber and H.W. Heldt, Annu. Rev. Plant Physiol. 32, 139-168 (1981).
3. S.C. Huber, Z. Pflanzenphysiol. 101, 49-54 (1981).
4. S.C. Huber, Plant Physiol. 71, 818-821 (1983).
5. N.J. Chatterton and J.E. Silvius, Plant Physiol. 64, 749-753 (1979).
6. N.J. Chatterton and J.E. Silvius, Physiol. Plant. 49, 141-144 (1980).
7. C. H. Foyer, S. Harbron, and D. Walker in: Photosynthesis IV. Regulation of carbon metabolism, G. Akoyunoglou, ed. (Balaban International Science Services, Philadelphia 1981) pp. 357-364.
8. S. Harbron, C. Foyer, and D. Walker, Arch. Biochem. Biophys. 212, 237-246 (1981).
9. S.C. Huber, T.W. Rufty, P.S. Kerr, and D.C. Doehlert, Curr. Topics in Plant Biochem. and Physiol. 2, 20-32.
10. R.T. Giaquinta, Plant Physiol. 61, 380-385 (1978).
11. J.C. Pollock, Plant Sci. Lett. 7, 27-31 (1976).
12. S.C. Huber, R.F. Wilson, and J.W. Burton, Plant Physiol. 73, 713-717 (1983).
13. P.S. Kerr, S.C. Huber, and D.W. Israel, Plant Physiol, in press (1984).
14. T.W. Rufty and S.C. Huber, Plant Physiol. 72, 474-480 (1983).
15. E. Van Schaftingen, L. Hue, and H.G. Hers, Biochem. J. 192, 897-901 (1980).
16. E. Furuya and K. Uyeda, Proc. Natl. Acad. Sci. 77, 5861-5864 (1980).
17. C. Cséke, N.F. Weeden, B.B. Buchanan, and K. Uyeda, Proc. Natl. Acad. Sci. 79, 4322-4326 (1982).
18. M. Stitt, G. Mieskes, H.D. Soling, and H.W. Heldt, FEBS Lett. 145, 217-222 (1982).
19. M. Stitt, R. Gerhardt, B. Kürzel, and H.W. Heldt, Plant Physiol. 72, 1139-1141 (1983).
20. S.C. Huber and D.M. Bickett, Plant Physiol. 74, 445-447 (1984).
21. J. Amir and J. Preiss, Plant Physiol. 69, 1027-1030 (1982).
22. D.C. Doehlert and S.C. Huber, FEBS Lett. 153, 293-297 (1983).
23. D.C. Doehlert and S.C. Huber, Plant Physiol. 73, 989-994 (1983).
24. G.L. Salerno and H.G. Pontis, FEBS Lett. 86, 263-267 (1978).
25. S.C. Huber, Z. Pflanzenphysiol. 102, 443-450 (1981).
26. T.W. Rufty, P.S. Kerr, and S.C. Huber, Plant Physiol. 73, 428-433 (1983).
27. P.S. Kerr, Ph.D. Thesis, North Carolina State University, Raleigh, NC (1984).
28. N. Terry and D.C. Mortimer, Can. J. Bot. 50, 1049-1054 (1972).
29. M. Stitt, W. Wirtz, and H.W. Heldt, Biochim. Biophys. Acta 593, 85-102 (1980).
30. C. Cséke and B.B. Buchanan, FEBS Lett. 155, 139-142 (1983).

Photosynthesis in Allopolyploid *Festuca*

D.D. Randall, C.J. Nelson, D.A. Sleper, C.D. Miles, C.F. Crane, R.W. Krueger, J.H.H. Wong, J.W. Poskuta

University of Missouri, Columbia, MO 65211 USA.

INTRODUCTION

Photosynthesis is responsible for the energy and carbon for plant growth, and yet most attempts to correlate increased photosynthesis per unit leaf area or weight with increased yield of crops, including tall fescue (1) have been unsuccessful. Photosynthesis and yields of C_4-plants are typically greater than those of C_3-plants, suggesting that increased photosynthesis should contribute to increased yields. Attempts to increase the efficiency of C_3-plant photosynthesis have included selections for decreased photorespiration [2], increased rates of apparent photosynthesis (APS) [3], and genetically engineering ribulose-bisphosphate carboxylase/oxygenase [4]. While there has been some progress in these pursuits, there has been limited success in translating these successes into increased plant yields. However, all of these efforts have continued to refine our knowledge of the photosynthetic process and its multiplicity of interacting molecular and physical processes.

This report describes our investigations on photosynthesis in cultivated and wild relatives of an allopolyploid C_3 forage grass, tall fescue, Festuca arundinacea , Schreb. Tall fescue belongs to the Bovinae section of the genus Festuca; tribe Festuceae. It evolved in the area surrounding the Mediterranean Sea and is found in its native state in Europe, Tunisia and the west and central area of Asia and Siberia [5]. The Festuca genus contains cultivated hexaploid and wild tetraploid, octoploid and decaploid varieties, all founded on a basic chromosome number of x=7 (Table I). A diploid species, F. pratensis Huds., morphologically resembles F. arundinacea and probably contributed a genome to some polyploid fescues. The phylogenetic relationships among tall fescue and its relatives are not completely understood and with few known diploids, phylogenetic analyses of the polyploids are difficult.

Table I. Proposed genomic formula of Festuca species.

Species	Somatic Number (2n)	Proposed Genomic Formula
F. pratensis Huds.	14	PP
F. arundinacea var. glaucescens Boiss	28	$G_1G_1G_2G_2$
F. mairei St. Yves	28	$M_1M_1M_2M_2$
F. arundinacea var. genuina Schreb.	42	$PPG_1G_1G_2G_2$
F. arundinacea var. atlantigena St. Yves	56	$G_1G_1G_2G_2M_1M_1M_2M_2$
F. arundinacea var. letourneuxiana and cirtensis	70	$QQG_1G_1G_2G_2M_1M_1M_2M_2$

Published 1985 by Elsevier Science Publishing Co., Inc.
Paul W. Ludden and John E. Burris, editors
NITROGEN FIXATION AND CO_2 METABOLISM

PHOTOSYNTHESIS AND RuBP CARBOXYLASE IN ALLOPOLYPLOID TALL FESCUE

Our original interest in the polyploidy aspect of this species was sparked by our discovery [3] that a particular genotype (I-16-2) had an APS that was 58 to 74% higher than 13 other genotypes [3,6]. Subsequently, Evans et al. [7] established that the high APS genotype was a decaploid (2n = 10x = 70) vs the other genotypes which were hexaploids (2n = 6x = 42). Table II illustrates typical results and also shows there were no differences in percentage photorespiration (PR) between the 10x and 6x genotypes. The obvious difference was the number of genomes or ploidy level of the genotypes. Thus the questions: Was gene dosage a factor? Could we see it expressed in RuBP Carboxylase (RuBPCase) levels?

Table II. Field measurements of APS of tall fescue[a].

A. Oct.-Nov.		mg CO_2 fixed$\cdot dm^{-2}\cdot h^{-1}$	
hexaploids (avg. of 10 genotypes)		21.6	
decaploid (I-16-2)		36.6	
B. June	**in air**	**in 1% O_2**	**% PR**
hexaploids	14.7(31.4)*	21.7(46.4)	32.5(32.2)
decaploid	39.4(57.9)	48.3(80.1)	27.7(27.7)

[a] Adapted from ref [6]. *(values) are mg CO_2 fixed $\cdot g^{-1}\cdot h^{-1}$

We extracted field or growth chamber grown tissue using six different methods and based the RuBPCase activity on either protein or Chl [6]. The ratio of RuBPCase activity for the decaploid: hexaploid ranged from 1.3 to 4.1 (average: 1.96). RuBPCase activity in isolated and ruptured chloroplasts exhibited a ratio of 1.6 (10x:6x) [6]. The ratios of activities of eight other enzymes were less than 1.0 for 10x:6x but the RuBPCase: RuBPOase ratio was unchanged [6]. RuBPCase purified from the high APS decaploid and a typical APS hexaploid was not different in physical properties or final specific activity [8]. The only difference was the amount of carboxylase present, 18 and 28% of soluble protein for the 6x and 10x genotypes, respectively. Thus, both APS and RuBPCase levels increased in the decaploid versus the hexaploids.

Were the different RuBPCase levels due to ploidy level *per se*? The literature on polyploidy (at that time) generally revealed that photosynthesis rates decreased with increased ploidy level [9,10,11,12,13], but with occasional increases [14]. Most studies utilized polyploid tissue derived by colchicine treatment or from matings utilizing unreduced gametes. Consequently they were "raw" polyploids which Stebbins [15] indicates are initially inferior. Polyploids generally exhibit increased cell size but frequently decreased cell number. The gigas character is prevalent in raw polyploids but "old" polyploids are frequently indistinguishable from diploids. Mesophyll cell volume and chloroplast number and size can be proportional to ploidy level [13,15,16,17]. Stomata density is usually decreased in polyploids but stomata are often larger and more open [10,11,18,19]. RuBPCase levels have only recently been examined as a function of ploidy level and will be discussed later.

With this lack of encouragement from the literature but intriguing data from our own studies, we selected genotypes solely on the basis of ploidy level for further study. Initially four genotypes at each of four ploidy levels (4x, 6x, 8x and 10x) were chosen [20], but later the diploid *F. pratensis* and tetraploid *F. mairei* St. Yves were included (Table I and III). All were naturally occurring allopolyploids. APS was measured by differential CO_2 analysis of attached, newly collared (mature) leaves using an open system [21]. Total (activated) RuBPCase activity was determined and RuBPCase proteins quantitated immunochemically (Table III) [20].

Table III. Relationship of APS characters among natural allopolyploids of Festuca growing in the field.[a]

Chrom No.	Genomes	APS*	SLW*	Sol.Prot	RuBPC**	RuBPC/prot
				---mg/g fr. wt.--		%
14	PP	20.8	5.1	42.2	6.8	16.7
28	$G_1G_1G_2G_2$	19.1	5.9	40.8	8.7	16.8
28	$M_1M_1M_2M_2$	28.7	8.7	32.5	9.6	30.8
42	PP $G_1G_1G_2G_2$	22.8	5.9	41.4	9.2	22.2
56	$G_1G_1G_2G_2M_1M_1M_2M_2$	30.1	5.7	37.2	10.1	26.2
70	QQ $G_1G_1G_2G_2M_1M_1M_2M_2$	32.2	5.6	37.4	10.5	28.4

[a]Mean of two experiments, APS*:mg CO_2 fixed·dm^{-2}·h^{-1},SLW*:mg·cm^{-2}.
**RuBPC:RuBP carboxylase, SLW:specific leaf weight.

Table IV. Influence of ploidy level on stomatal density on the upper surface, leaf diffusive resistance, and APS of tall fescue leaf blades. Adapted from ref [22].

Ploidy	Genomes	Stomata Density	Stomata Length	Diff. Res. Up	Diff. Res. Low	APS mg CO_2·h^{-1}		Cell Vol.*
		No·mm^{-2}	um	s·cm^{-1}		g^{-1}	dm^{-2}	
4x	$G_1G_1G_2G_2$	43.6	41.8	6.7	17.7	32.7	15.9	1.0
6x	$PPG_1G_1G_2G_2$	36.3	45.5	2.4	5.5	34.6	18.8	6.0
8x	$G_1G_1G_2G_2M_1M_1M_2M_2$	30.6	45.8	2.4	8.8	46.6	22.3	7.2
10x	$QQG_1G_1G_2G_2M_1M_1M_2M_2$	47.2	50.7	0.8	3.7	44.4	21.7	4.7

*Cell volume: Volume index assumes 3rd dimension was proportional to width and 4x was equal to 1.0.

As in many studies stomata density decreased and length increased as ploidy increased from 4x to 8x. Stomata were highest in density and longest in the 10x. Diffusive resistance decreased with increasing genome number, and calculated cell volume increased from 4x to 8x while at 10x it was between 4x and 6x. The number of veins in the tissue tended to decrease with ploidy level and probably reflected leaf width since cell numbers between veins were generally similar [22].

These studies provide supportive evidence that APS was increased by higher levels of RuBPCase as suggested by others [23,24,25]. The correlation between APS and concentration of RuBPCase supports the hypothesis that polyploidization in tall fescue could lead to an increase in the percentage of RuBPCase present, which in turn contributes at least in part to the increase in net photosynthesis in polyploid Festuca. Dean and Leech [26] recently reported little change in percentage of RuBPCase of soluble protein for allopolyploid wheat but a 26% increase in RuBPCase per leaf section and a constant RuBPCase/cell: nDNA/cell ratio. Studies with isogenic polyploid alfalfa showed RuBPCase levels were similar when based on a protein or Chl basis [27,28]. When RuBPCase levels were determined per protoplast the tetraploid alfalfa protoplasts contained about twice the RuBPCase as the diploid alfalfa [28]. This was a consequence of cell volume being about twice as large.

The observed correlation between APS and RuBPCase content in allopolyploid Festuca [20] is not sufficient to establish that the increase in APS is due to the increase in the carboxylating enzyme. However, it is logical that an increase in content of the primary carboxylating enzyme of photosynthesis could contribute to the increase in APS observed and could help overcome the trend of decreasing stomatal density associated with ploidy. The increased RuBPCase with ploidy may be associated with increased number of chloroplasts associated with large cell size, but the cell size of the 10x was intermediate between the 4x and 6x. The increased RuBPcase may also reflect that the regulation of RuBPCase synthesis is controlled by synthesis of the small subunit (SSU) which is coded by nuclear DNA. Therefore, more SSU synthesis due to increased genome number would result in more holoenzyme [26,29,30]. The high APS for our decaploids could, in part, be due to the high stomatal density associated with small cell size and low diffusive resistance (Table IV) [22] and/or alternatively to a more favorable in vivo activation level of RuBPCase [31]. In summary, our examination of anatomical features of this series revealed little to explain the high APS and RuBPCase levels. Wilson and Cooper [32] have previously shown with ryegrass that high APS was associated with small cell volume, however tall fescue does not necessarily follow this pattern. Parker and Ford [33] have suggested that the amount of mesophyll surface exposed to air space is more critical than the surface to volume ratio of the cells. Jellings and Leech [34] examined 20 leaf characters of nine Triticum genotypes at three ploidy levels and found that ploidy level influenced only cell size, chloroplast number and leaf thickness in allopolyploid wheat. They reported that photosynthetic capacity was negatively correlated with the ratio of mesophyll cell size to nuclear genome size [34].

PHOTOSYNTHETIC ELECTRON TRANSPORT AND ATP SYNTHESIS

Could the photosynthetic electron transport and phosphorylation properties of these genotypes explain their relative APS rates? An examination of photosynthetic electron transport and photophosphorylation in isolated tall fescue chloroplasts showed that these processes exhibited high rates that were comparable to those of spinach chloroplasts [35]. The use of isolated chloroplasts also avoided problems with leaf morphology. Rates of electron transport and ATP formation were compared in isolated chloroplasts from the expanded 24 genotype, allopolyploid series (2x to 10x) (Table I). No correlation between ploidy level and photosynthetic electron transport activity was found. Figure I illustrates these results for PSI electron transport [36]. The original high APS decaploid (I-16-2, genotype "m", Fig. 1) consistently exhibited high

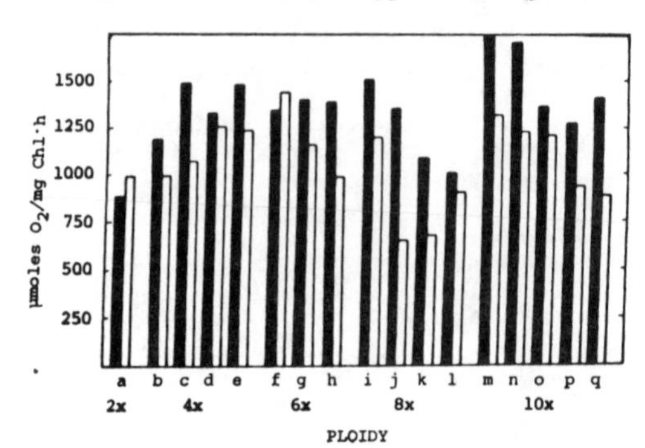

Figure I. PSI electron transport of isolated chloroplasts from 24 genotypes of Festuca grown under greenhouse (▮) and field (▯) conditions. From ref [36] with permission.

PSI-mediated electron flow [37]. Further characterization of PSI (Table V) in this decaploid showed increased P700/Chl levels, increased plastocyanin with increased charge density (more negative pI) [37], an increased potential for cyclic electron flow and phosphorylation (Table V) and no change in Chl a/b ratio compared with the hexaploid. These studies gave no support for the hypothesis that gene dosage or polyploidy *per se* was the reason for the differences in APS rates. But the results show that particular genotypes exhibited unique and supportive electron flow properties conducive to high APS rates (e.g. I-16-2 = m, *F. mairei* = e, Fig. 1).

Table V. Comparison of PSI characteristics.

Ploidy level	Electron Transport	ATP Syn.	P:2e⁻	Chl/P700	Plasto-cyanin	Chl/PC
	μmol O_2 / mg Chl·h	μmol Pi / mg Chl·h	ratio	ratio	$\mu g \cdot mm^{-2}$	ratio
6x[a]	531	292	0.59	417	0.025	157
10x	614	427	0.79	343	0.035	117

[a]6x = V6-802; 10x = I-16-2; Adapted from ref [36,37].

ASSIMILATION PRODUCTS AND PHOTOSYNTHATE TRANSLOCATION

An early observation with the high APS decaploid in comparison with typical hexaploids was that smaller "starch pellets" occurred in leaf extracts of the decaploid, and the decaploid also had less leaf area. Where is all the carbon that is being fixed by this decaploid? Could the assimilation products shed light on the high APS of the decaploid or the polyploidy APS pattern? We found that APS of the decaploid saturated at higher light intensity than the hexaploid but that the products of newly assimilated $^{14}CO_2$ were the same [38]. Quantitative differences were found in the amino acids (higher in 10X) while both genotypes exhibited large glycerate pools (1% dry wt) and low serine and glycine pools [38]. Both genotypes had less than 1% of total radioactivity in starch after 10 min.

Sucrose was the major product of CO_2 assimilation. Rates of sucrose synthesis were 7.76 and 4.94 $\mu mol \cdot g$ fr $wt^{-1} \cdot h^{-1}$ for the high APS decaploid and typical hexaploid, respectively, but the total sucrose pool was 23% greater and the free fructose pool was 3 times greater in the hexaploid [38]. In fact, the water-soluble carbohydrates (WSC) were 33% greater in the hexaploid [38]. Tall fescue accumulates fructans, but leaf fructan levels were low and about equal to those of starch (ca. 1-2% dry wt with the fructan concentration in 6x > 10x) [38]. Both genotypes have the leaf starch biosynthetic capacity, since $^{14}CO_2$ fixation (in 1.1% CO_2) experiments with detached leaf segments yielded 10% of the label in the starch fraction [38].

The sucrose level in the leaf is the result of synthetic, degradative and translocation processes. We examined the synthetic and degradative enzyme activities of the high APS decaploid compared to the hexaploid (Table VI). It is generally believed that sucrose-P-synthetase (SPS) is responsible for synthesis [39] and sucrose synthetase (SS) is involved in UDPGlc formation from sucrose. SPS is usually present at considerably greater levels than SS in leaf tissue [39]. However, SS activity was essentially the same as SPS activity (Table VI). In fact the high levels of SS activity in tall fescue leaves may suggest a synthetic role for the enzyme in tall fescue, especially when one considers the large free fructose pool [38]. However, neither SPS nor SS levels appears to explain the difference in sucrose synthesis rates.

Table VI. Enzymes related to sucrose synthesis.

Enzyme		Decaploid	Hexaploid
		($\mu mol \cdot h^{-1} \cdot mg\ prot^{-1}$)	
Sucrose-P Synthetase		0.82	0.98
Sucrose Synthetase		0.84	0.69
Invertase	pH 5.0	0.051	0.041
	pH 7.3	0.019	0.023
Fructose-6-P Phosphatase		0	0

10x = I-16-2; 6x = V6-802

We used $^{14}CO_2$ fixation and pulse-chase experiments to show that translocation of photosynthate from a fed portion of a leaf was slow (6 to 10 cm h^{-1}) in tall fescue, but that over 24 h the decaploid translocated about 1.5 fold more ^{14}C than the hexaploid (Table VII). Thus, the 10x translocated 1.5 fold more ^{14}C and fixed $^{14}CO_2$ at ca. 1.6 fold greater rate resulting in 2.4 fold more carbon translocated. The relative distribution 24h after the $^{14}CO_2$ pulse was 40, 33 and 27% in the fed leaf, stembase and roots, respectively, for the decaploid, and 90, 3 and 2% in the fed leaf, stembase and roots, respectively, for the hexaploid. The difference in translocation and partitioning could not be explained by vein number or distribution, i.e. the vein number is less, but the leaf is narrower and the distance between the veins is greater for the decaploid [22 and J.H.H. Wong unpublished].

Table VII. Percentage of assimilate translocated 6 and 24 h after a 10 min pulse of $^{14}CO_2$ at steady state photosynthesis.

Ploidy level	Intact[a]		Simplified	
	6h(6L)	24h(6L·12D·6L)	6h(6L)	24h(6L·12D·6L)
10x	40%	79%	44%	86%
6x	26%	48%	24%	69%
ratio (10x/6x)	1.53	1.62	1.84	1.24

[a]L, D. = in light and darkness, respectively, after labeling. Intact: all leaves present; simplified: fed leaf and tiller only.

Does the increased APS rate of the decaploid drive increased sucrose synthesis and subsequently increase translocation? Or does increased synthesis and translocation (smaller pool size) result in less feedback on photosynthesis in the decaploid, or does the larger pool size and decreased translocation rate feed back and decrease the assimilation rate in the hexaploid? The third alternative is that there is no relationship between sucrose levels and APS rates. In theory the difference in sucrose content should result from, rather than cause, the photosynthetic rate differential.

PHOTOSYNTHESIS AND YIELD RELATIONSHIPS

Progeny resulting from 6x by 10x matings that included the high APS decaploid were used to evaluate yield responses from wide crosses. Field and growth chamber experiments using serial harvest techniques verified that the 6x parent partitioned more dry matter to top growth and less to root growth and storage in the stem bases, Table VIII [40]. During the 8-week regrowth period leaf area ratios averaged 0.23 $dm^2 \cdot g^{-1}$ (leaf area/plant weight) for the decaploid, 0.38 $dm^2 \cdot g^{-1}$ for the hexaploid, and 0.34 $dm^2 \cdot g^{-1}$ for the hybrids. This was also reflected by the more rapid leaf elongation rate of the hexaploid genotype and hybrids versus the decaploid. The increase in leaf weight was similar for the hybrids and the hexaploid, whereas the increase in the stembase weight was similar for the decaploid and the hybrids [Table VIII]. Root weight increase of the hybrids was intermediate. Total increase in plant weight was similar for the hexaploid and the decaploid, but heterosis was evident for the hybrids. Hybrids had an intermediate APS rate. Hybrid vigor occurred since hybrids inherited in a dominant manner the increased use of photosynthate for both leaves (from the hexaploid) and stem bases (from the decaploid). The multiplicative dominance-by-dominance type of gene action for leaf and stembase weight was expressed as heterosis for total plant weight increase.

Table VIII. Increase in plant weights during an 8-week growth period [40], (decaploid = I-16-2, hexaploid =V2-29, hybrids = mean of 13).

	Decaploid	Hexaploid	Hybrids
		$g \cdot pot^{-1}$	
Leaves	15.3	26.7	27.7
Stembases	12.1	1.9	11.0
Roots	7.0	4.6	5.7
Total	34.4	33.2	44.4

Carbon storage was also influenced as the concentration of WSC reflects the difference between APS and growth. At the end of the regrowth period, the decaploid had more WSC available in the stembases (372 $mg \cdot pot^{-1}$) than did the hexaploid (229 $mg \cdot pot^{-1}$), but the hybrids had the most (474 $mg \cdot pot^{-1}$). Again this reflects the increased photosynthetic capacity of the hybrids over the hexaploid, as well as an inadequate sink strength to fully utilize the photosynthate for leaf growth. Selection for rapid leaf growth and improved sink strength in *Festuca* has been achieved with other genetic materials [41].

Since the data suggested that leaf growth rates were inversely related to leaf photosynthesis, we sought to determine the cause:effect relationship. Ramets of well developed tillers of the high APS decaploid (I-16-2) and a typical APS hexaploid (V-2-29) were cut 7-cm in length and were immersed for 24 h in solutions of 0, 100, and 500 ppm gibberellic acid (GA_3). Plants were allowed to grow for 3 weeks at 20°C in hydroponic culture [42]. The hexaploid had a faster leaf elongation rate (LER), wider leaves (LW), and faster leaf area expansion rate (LAER) (Table IX). Application of 100 ppm GA increased LER of both genotypes by ca. 90%, but decreased LW slightly. The net effect was a 60 to 80% increase in LAER. SLW was similar for both genotypes at 0 GA, but GA treatment decreased SLW in the hexaploid, but increased SLW in the decaploid.

Table IX. Leaf growth parameters and gas exchange characteristics of tall fescue genotypes treated with gibberellic acid (GA_3) [42].

GEN	GA	LER	LW	LAER	SLW	APS	APS/SLW	MES[1]	CSA[2]
	ppm	$mm \cdot d^{-1}$	mm	$mm^2 \cdot d^{-1}$	$mg \cdot cm^{-2}$	$mg \cdot dm^{-2} \cdot h^{-1}$		$no. \cdot UA^{-1}$	$mm^2 \cdot mm^{-3}$
Hex	0	33.1	10.2	357	7.8	26.5	3.39	1.46	22.4
	100	63.0	8.8	565	7.1	19.4	2.75	1.28	33.8
	500	67.0	8.0	567	6.4	12.4	1.93	1.28	30.6
Dec.	0	24.0	6.3	157	8.4	46.4	5.52	1.26	25.1
	100	45.9	5.8	284	8.9	39.7	4.47	1.27	34.7
	500	48.1	5.4	269	10.6	34.5	3.25	1.59	30.5

[1]Cells/unit area; [2]Cell surface area/volume of leaf blade [43].

The APS of the decaploid, as usual, was 75% higher than the hexaploid at 0 GA. When LAER was increased by adding 100 ppm GA, the APS decrease was proportionately much less than the LAER increase. This shows that LAER, which is closely related to final leaf size [41], and APS can be influenced independently. This is consistent with our earlier research where genetic selection for a 2-fold difference between populations in LAER caused less than a 10% difference in APS per unit leaf area [41].

The number of mesophyll cells per unit leaf area was slightly higher for the decaploid than for the hexaploid at 0 GA, but cell surface areas tended to be similar. Cell shape for tall fescue is not heavily convoluted as described for wheat [33], but tends to be oblate spheroid, spheres flattened at the poles. Cell number was decreased by GA in the hexaploid, but was increased in the decaploid. Cell surface area per unit leaf volume was increased by GA. Our earlier work [44,45] had shown that rapid leaf elongation for hexaploids was associated with both more cells and larger cells but APS was basically not influenced. With GA (Table IX) a 50% increase in cell surface area was associated with a 27% decrease in APS for the hexaploid, and a 38% increase in cell surface area was associated with a 14% decrease in APS for the decaploid [42].

GENOMIC INFLUENCES ON APS AND CARBOXYLASE

The question then remains, why does the decaploid have a higher APS than the hexaploid? Some insight can be gained by reviewing the genomic relationships of the ploidy series (Table I). Our earlier research showed an increase in APS as ploidy increased from 4x (*F.a.* var. *glaucescens*) to 10x [20,22] which is also evident in Table III. It is also clear that *F. mairei* contributes four genomes to the 8x and 10x genotypes [48]. There is evidence that *F. mairei* is an important contributor to the APS potential of those ploidy levels (Table III). *F.mairei* has a low concentration of soluble protein compared with the other genotypes, probably because it has a high SLW as the leaf visually appeared to be higher in structural tissue. Despite the low soluble protein of *F. mairei* the RuBPCase content is relatively high which gives this species a high RuBPCase/protein ratio. This is consistent with our earlier data [6,20] that one of the factors contributing to the genotype (ploidy) differences is the proportion of total protein made up of RuBPCase. When *F. mairei* and *F. a. glaucescens* were crossed, the progeny were intermediate in APS suggesting additive gene action [J.W. Poskuta, unpublished]. The proportion of RuBPCase was changed yet the *in vivo* RuBPCase/Oase ratio was unchanged. These 4x progeny have essentially the same genomic composition as the polyhaploids from the 10x with the exclusion of the Q genome. Further, polyhaploids of the decaploid had the same *in vivo* carboxylase/oxygenase ratio as the decaploid. The haploid plants were less vigorous and smaller than other *Festucas* tested [J.W. Poskuta, unpublished]. They need to be evaluated further as some haploids had APS rates similar to the decaploid parent.

It is interesting to look even further at the genetic control of photosynthesis, particularly as it relates to genomic structure of the various *Festuca* species examined. Chandrasekharan and Thomas [46,47] proposed that the decaploid contains four genomes each from *F. mairei* and *F. a.* var. *glaucescens*, but with two genomes from the diploid *F. pratensis* Huds (Table I). Our recent evidence [48] indicates that the *F. pratensis* genome is not part of the decaploid's genomic structure but rather, some other unknown diploid (QQ) is suggested. The F_1 hybrids from the tetraploid crosses are nearly identical genomically to the polyhaploids from the decaploid with the exception of the genome (QQ) from the unknown diploid. *F. mairei* has an APS rate equivalent to that of the decaploid and consistently shares high RuBPCase levels, high PSI mediated electron transport rates and high plastocyanin levels in common with the high APS decaploid [R.W. Krueger, unpublished].

CONCLUSIONS

Materials used in this research allow us to speculate more conclusively on photosynthesis and its relationship to gene action and polyploidy *per se* in *Festuca* spp. The intermediate rates of APS and PR in tetraploid F_1 hybrids of *F. a.* var. *glaucesens* and *F. mairei* versus the parents suggests that inheritance is largely due to additive gene action [3]. This supports earlier work showing that the inheritance of APS on a leaf area basis of hexaploid tall fescue genotypes was largely due to additive genetic variance [3].

It appears the increased photosynthesis of the decaploid is largely the result of obtaining superior genes from the *F. mairei* genomes (Table III). Ploidy *per se* may not be the most likely explanation of increased photosynthetic activity in *Festuca* spp. The increased number of nuclear genomes could result in increased expression of RuBPCase SSU which consequently controls large subunit formation and increases the holoenzyme in the chloroplasts. Seemann and Berry [25] support the conclusion that APS is proportional to RuBPCase at fixed internal CO_2 concentrations. Others have also reported that APS is proportional to RuBPCase levels [23,24]. This adds further support to our hypothesis that additive gene action is the predominant mode of inheritance for photosynthesis. If increasing ploidy *per se* was entirely responsible for increased photosynthesis, we would expect the resulting polyhaploids to be more uniformly lower in APS than the decaploid and this does not appear to be the case.

Our data [6,20] (Table III) have shown an increase in APS as ploidy level increases in *Festuca* from the 4x to the 8x level, and a small increase between the 8x and 10x levels. It is possible that the frequency of desirable alleles (i.e. increased PSI,cyclic phosphorylation, translocation) controlling photosynthesis processes other than RuBPCase have also been increased as the result of adding genomes. It is likely that in *Festuca* the increase in frequency of desirable alleles for photosynthesis is facilitated by increasing ploidy levels, but more importantly, natural selection had likely occurred for increased photosynthetic activity of progenitors before extensive polyploidization had developed to the decaploid level.

ACKNOWLEDGMENTS: This research was supported by USDA/SEA CRGO Grant 79-59-2291-1-1-366-1 and the Missouri Agricultural Experiment Station.

REFERENCES

1. Nelson, C.J., K.H. Asay and G.L. Horst, Crop Sci. 15,476-479 (1975).
2. Zelitch, I. and P.R. Day, Plant Physiol. 52, 33-37 (1973).
3. Asay, K.H., C.J. Nelson and G.L. Horst, Crop Sci. 14, 571-574 (1974).
4. McIntosh, L., J. Hirschberg, C. Somerville and John Fitchen, Advances in Photosynthesis Research, C. Sybesma ed. IV, 483-490 (Nijhoff/Junk Publishers, The Hague 1984).
5. Terrell, E.E., *In*: Tall Fescue, R.C. Buckner and L.P. Bush eds. (Amer. Soc. of Agron., Madison, Wisc. 1979). p. 31-39.
6. Randall, D.D., C.J. Nelson and K.H. Asay, Plant Physiol. 59 38-41 (1977).
7. Evans, G.M., K.H. Asay, and R.A. Jenkins, Crop Sci. 13, 376-379 (1973).
8. Leimkuhler, W., MS Thesis, University of Missouri-Columbia (1978).

9. Ekdahl, I., Ark. Bot. 31A, 1-48 (1944).
10. Bjuram, B., Physiol. Plant 12, 183-197 (1959).
11. Frydrya, J., Photosynthetica 4, 139-145 (1970).
12. Austin, R.B., C.L. Morgan, M.A. Ford and S.G. Bhagwat, Ann. Bot. 49, 127-189 (1982).
13. Cukrova, V. and N. Auratovscukova, Photosynthetica 2, 227-237 (1968).
14. Beysel, D., Der Zuchter 27, 261-277 (1957).
15. Stebbins, E.G., Chromosomal Evolution in Higher Plants (Addison-Wesley Pub. Co. Reading, Mass. 1971) p. 125-154.
16. Butterfass, T., Patterns of Chloroplast Reproduction (Springer-Verlag, New York 1979). pp. 204.
17. DeMaggio. A.E. and D.A. Settler, Exp. Cell Res. 67, 287-294 (1971).
18. DeMaggio, A.E., Bioscience 21, 313-316 (1971).
19. Tal, M. and T. Gardi, Physiol. Plant 38, 257-261 (1976).
20. Joseph, M.C., D.D. Randall and C.J. Nelson, Plant Physiol. 68, 894-898 (1981).
21. Nelson, C.J., K.H. Asay, G.L. Horst and E.S. Hilderbrand, Crop Sci. 14, 26-28 (1974).
22. Bryne, M.C., C.J. Nelson, and D.D. Randall, Plant Physiol. 68, 891-893 (1981).
23. Wareing, P.F., M.M. Khalifa, and K.J. Treharne, Nature 220, 453-457 (1968).
24. Bjorkman, O., Physiol. Plant. 21, 84-89 (1968).
25. Seemann, J.R. and J.A. Berry, Carnegie Inst. Wash Yearbook 81, 78-83 (1982).
26. Dean, C. and R.M. Leech, Plant Physiol. 70, 1605-1608 (1982).
27. Meyers, S.P., S.L. Nichols, G.R. Baer, W.T. Molin and L.E. Schrader, Plant Physiol. 70, 1704-1709 (1982).
28. Molin, W.T., S.P. Meyers, G.R. Baer and L.E. Schrader, Plant Physiol. 70, 1710-1714 (1982).
29. Barraclough, R. and R.J. Ellis, Eur. J. Biochem. 94, 165-177 (1979).
30. Dean, C. and R.M. Leech, FEBS Lett. 140, 113-116 (1982).
31. Lorimer, G.H., Ann. Rev. Plant Physiol. 32, 349-383 (1981).
32. Wilson, D. and J.P. Cooper, New Phytol. 69, 1838-1844 (1970).
33. Parker, M.L. and M.A. Ford, Ann. Bot. 49, 165-176 (1982).
34. Jellings, A.J. and R.M. Leech, New Phytol. 96, 371-382 (1984).
35. Krueger, R.W. and C.D. Miles, Plant Physiol. 67, 763-767 (1981).
36. Krueger, R.W. and C.D. Miles, Plant Physiol. 68, 1110-1114 (1981).
37. Krueger, R.W., C.D. Miles and D.D. Randall, Plant Physiol., In Press (1984).
38. Wong, J.H.H., D.D. Randall and C.J. Nelson, Plant Physiol. 72, 16-21 (1983).
39. Akazawa, T. and K. Okamoto, The Biochemistry of Plants, 3, 199-220 J. Preiss ed. (Academic Press, New York 1980).
40. Nelson, C.J., K.M. Zarrough, D.D. Randall, and D.A. Sleper, Curr. Topics Plant Biochem. Physiol. 1, 233 (1983).
41. Nelson, C.J. and D.A. Sleper, In: J.A. Smith and V.W. Hays (ed) Proc. XIV Int. Grassld. Conf. (Lexington, KY, 1983) p. 413-416.
42. Poskuta, J.W., C.J. Nelson, J.H. Coutts and T.L. Vassey, Curr. Topics Plant Biochem. Physiol. 3, 174 (1984).
43. Thain, J.F., J. Exp. Bot. 34, 87-94 (1983).
44. Volenec, J.J., and C.J. Nelson, Crop Sci. 23, 720-724 (1983).
45. Volenec, J.J., and C.J. Nelson, Crop Sci. 21, 381-385 (1981).
46. Chandrasekharan, P. and H. Thomas, Z. Pflanzenzucht. 65, 345-354 (1971).
47. Chandrasekharan, P. and H. Thomas, Z. Pflanzenzucht. 66, 76-80 (1971).
48. Crane, C.F., K.L. Hunt and D.A. Sleper, Can. J. Genetics and Cytol. In Press (1984).

PUBLICATIONS BY R. H. BURRIS AND CO-WORKERS

1. K. W. Franke, R. H. Burris and R. S. Hutton. A New Colorimetric Procedure Adapted to Selenium Determination. Ind. Eng. Chem. Analytical Edition 8, 435 (1936).
2. W. W. Umbreit and R. H. Burris. Composition of Soybean Nodules and Root Nodule Bacteria. Soil Sci. 45, 111-126 (1938).
3. O. Wyss, R. H. Burris and P. W. Wilson. Occurrence and Significance of Oxalacetic Acid in Plant Tissues. Proc. Soc. Exp. Biol. Med. 40, 372-375 (1939).
4. R. H. Burris and P. W. Wilson. Respiratory Enzyme Systems in Symbiotic Nitrogen Fixation. Cold Spring Harbor Symposia on Quant. Biol. 7, 349-361 (1939).
5. D. W. Thorne and R. H. Burris. Respiratory Enzyme Systems in Symbiotic Nitrogen Fixation. II. The Respiration of Rhizobium from Legume Nodules and Laboratory Cultures. J. Bacteriol. 39, 187-196 (1940).
6. R. H. Burris and P. W. Wilson. Measures of Respiratory Activity with Resting Cells. Proc. Soc. Exp. Biol. Med. 45, 721-726 (1940).
7. R. H. Burris. Chapter 5 in "Respiratory Enzymes," (Elvehjem and Wilson, *et al*., eds.), Burgess Publishing Co., Minneapolis Minn., 1939, pp. 93-101.
8. R. H. Burris and C. E. Miller. Application of N^{15} to the Study of Biological Nitrogen Fixation. Science 93, 114-115 (1941).
9. D. Burk and R. H. Burris. Biochemical Nitrogen Fixation. Annual Review of Biochemistry 10, 587-618 (1941).
10. R. H. Burris. Failure of Barley to Fix Molecular N^{15}. Science 94, 238-239 (1941).
11. R. H. Burris. Criteria for Experiments with Isotopes. A Symposium on Respiratory Enzymes, p. 256 (1941).
12. R. H. Burris and P. W. Wilson. Liebig and the Microbiologist. Amer. Fertilizer 94, 9-11, 24, 26 (1941).
13. R. H. Burris. Distribution of Isotopic Nitrogen in *Azotobacter vinelandii*, J. Biol. Chem. 143, 509-517 (1942).
14. R. H. Burris and P. W. Wilson. Oxidation and Assimilation of Glucose by the Root Nodule Bacteria. J. Cell. and Comp. Physiol. 19, 361-371 (1942).
15. S. B. Lee, R. H. Burris and P. W. Wilson. Cell-free Enzymes of *Azotobacter vinelandii*. Proc. Soc. Exp. Biol. Med. 50, 96-98 (1942).
16. S. B. Lee and R. H. Burris. Large-scale Production of Azotobacter. Ind. Eng. Chem. 35, 354-357 (1943).
17. P. W. Wilson, R. H. Burris and W. B. Coffee. Hydrogenase and Symbiotic Nitrogen Fixation. J. Biol. Chem. 147, 475-481 (1943).
18. R. H. Burris, F. J. Eppling, H. B. Wahlin and P. W. Wilson. Detection of Nitrogen Fixation with Isotopic Nitrogen. J. Biol. Chem. 148, 349-357 (1943).
19. P. W. Wilson, J. F. Hull and R. H. Burris. Competition between Free and Combined Nitrogen in Nutrition of Azotobacter. Proc. Nat. Acad. Sci. 29, 289-294 (1943).
20. P. W. Wilson, R. H. Burris and C. J. Lind. The Dissociation Constant in Nitrogen Fixation by Azotobacter. Proc. Nat. Acad. Sci. 28, 243-250 (1942).
21. R. H. Burris, A. S. Phelps and J. B. Wilson. Adaptations of Rhizobium and Azotobacter. Soil Sci. Soc. Amer. Proc. 7, 272-275 (1942).
22. R. H. Burris, F. J. Eppling, H. B. Wahlin and P. W. Wilson. Studies of Biological Nitrogen Fixation with Isotopic Nitrogen. Soil Sci. Soc. Amer. Proc. 7, 258-262 (1942).
23. R. H. Burris and E. Haas. The Red Pigment of Leguminous Root Nodules. J. Biol. Chem. 155, 227-229 (1944).

24. W. W. Umbreit, R. H. Burris and J. F. Stauffer. Manometric Techniques and Related Methods for the Study of Tissue Metabolism. Burgess Publishing Co., Minneapolis, Minn. (1945).
25. R. H. Burris and P. W. Wilson. Biological Nitrogen Fixation. Annual Review of Biochemistry 14, 685-708 (1945).
26. H. Koffler, R. L. Emerson, D. Perlman and R. H. Burris. Chemical Changes in Submerged Penicillin Fermentations. J. Bacteriol. 50, 517-548 (1945).
27. H. Koffler, S. G. Knight, R. L. Emerson and R. H. Burris. The Effect of Certain Chemicals on Penicillin Production and Mold Metabolism in Shake Flask Fermentations. J. Bacteriol. 50, 549-559 (1945).
28. H. Koffler, S. G. Knight, W. C. Frazier and R. H. Burris. Metabolic Changes in Submerged Penicillin Fermentations on Synthetic Media. J. Bacteriol. 51, 385-392 (1946).
29. R. H. Burris and P. W. Wilson. Comparison of the Metabolism of Ammonia and Molecular Nitrogen in Azotobacter. J. Biol. Chem. 165, 595-598 (1946).
30. R. H. Burris and P. W. Wilson. Ammonia as an Intermediate in Nitrogen Fixation by Azotobacter. J. Bacteriol. 52, 505-512 (1946).
31. R. H. Burris and P. W. Wilson. Characteristics of the Nitrogen-fixing Enzyme System in Nostoc muscorum. Bot. Gaz. 108, 254-262 (1946).
32. O. E. Olson, R. H. Burris and C. A. Elvehjem. A Preliminary Report of the "Folic Acid" Content of Certain Foods. J. Amer. Dietetic Assn. 23, 200-203 (1947).
33. H. N. Little and R. H. Burris. Activity of the Red Pigment from Leguminous Root Nodules. J. Amer. Chem. Soc. 69, 838-841 (1947).
34. P. W. Wilson and R. H. Burris. The Mechanism of Biological Nitrogen Fixation. Bacteriol. Rev. 11, 41-73 (1947).
35. M. J. Kreko Leonard and R. H. Burris. A Survey of Transaminases in Plants. J. Biol. Chem. 170, 701-709 (1947).
36. H. A. Machata, R. H. Burris and P. W. Wilson. Fixation of Isotopic Nitrogen by Excised Nodules. J. Biol. Chem. 171, 605-609 (1947).
37. D. M. Powelson, P. W. Wilson and R. H. Burris. Oxidation of Glucose, Glycerol, and Acetate by Staphylococcus aureus. Biochem. J. 41, 486-491 (1947).
38. R. MacVicar and R. H. Burris. The Relation of Boron to Certain Plant Oxidases. Arch. Biochem. 17, 31-39 (1948).
39. O. E. Olson, E. E. C. Fager, R. H. Burris and C. A. Elvehjem. Folic Acid Activity in Homogenates of Rat Liver. J. Biol. Chem. 174, 319-326 (1948).
40. D. M. Molnar, R. H. Burris and P. W. Wilson. The Effect of Various Gases on Nitrogen Fixation by Azotobacter. J. Amer. Chem. Soc. 70, 1713-1716 (1948).
41. R. E. Stutz and R. H. Burris. Factors Influencing Oil Content of Potato Chips. Food Industries 20, 1146-1149, 1243-1246 (1948) (August).
42. O. E. Olson, E. E. C. Fager, R. H. Burris and C. A. Elvehjem. The Use of a Hog Kidney Conjugase in the Assay of Plant Materials for Folic Acid. Arch. Biochem. 18, 261-270 (1948).
43. R. MacVicar and R. H. Burris. Translocation Studies in Tomato Using Ammonium Sulfate Labeled with N^{15}. Amer. J. Bot. 35, 567-570 (1948).
44. R. MacVicar and R. H. Burris. Studies on Nitrogen Metabolism in Tomato with Use of Isotopically Labeled Ammonium Sulfate. J. Biol. Chem. 176, 511-516 (1948).
45. H. A. Lardy, W. E. Gilson, J. Hipple and R. H. Burris. Circular Bath and Shaking Mechanism for Manometric Microapparatus. Anal. Chem. 20, 1100-1102 (1948).
46. E. E. C. Fager, O. E. Olson, R. H. Burris and C. A. Elvehjem. Folic Acid in Vegetables and Certain Other Plant Materials. Food Res. 14, 1-8 (1949).

47. J. E. Mitchell, R. H. Burris and A. J. Riker. Inhibition of Respiration in Plant Tissues by Callus Stimulating Substances and Related Chemicals. Amer. J. Bot. 36, 368-378 (1949).

48. C. O. Clagett, N. E. Tolbert and R. H. Burris. Oxidation of α-Hydroxy Acids by Enzymes from Plants. J. Biol. Chem. 178, 977-987 (1949).

49. H. A. Lardy, R. L. Potter and R. H. Burris. Metabolic Functions of Biotin. I. The Role of Biotin in Bicarbonate Utilization by *Lactobacillus arabinosus* Studied with C^{14}. J. Biol. Chem. 179, 721-731 (1949).

50. E. S. Lindstrom, R. H. Burris and P. W. Wilson. Nitrogen Fixation by Photosynthetic Bacteria. J. Bacteriol. 58, 313-316 (1949).

51. N. E. Tolbert, C. O. Clagett and R. H. Burris. Products of the Oxidation of Glycolic Acid and l-Lactic Acid by Enzymes from Tobacco Leaves. J. Biol. Chem. 181, 905-914 (1949).

52. R. H. Burris, P. W. Wilson and R. E. Stutz. Incorporation of Isotopic Carbon into Compounds by Biosynthesis. Bot. Gaz. 111, 63-69 (1949).

53. A. C. Wagenknecht and R. H. Burris. Indoleacetic Acid Inactivating Enzymes from Bean Roots and Pea Seedlings. Arch. Biochem. 25, 30-53 (1950).

54. R. H. Burris. Isotopes as Tracers in Plants. Bot. Rev. 16, 150-180 (1950).

55. N. E. Tolbert and R. H. Burris. Light Activation of the Plant Enzyme which Oxidizes Glycolic Acid. J. Biol. Chem. 186, 791-804 (1950).

56. B. P. Ghosh and R. H. Burris. Utilization of Nitrogenous Compounds by Plants. Soil Sci. 70, 187-203 (1950).

57. R. E. Stutz and R. H. Burris. Photosynthesis and Metabolism of Organic Acids in Higher Plants. Plant Physiol. 26, 226-243 (1951).

58. R. H. Burris. Radioisotopes in Biology. The Science Counselor 14, 52-53 (1951).

59. I. Zelitch, E. D. Rosenblum, R. H. Burris and P. W. Wilson. Isolation of the Key Intermediate in Biological Nitrogen Fixation by Clostridium. J. Biol. Chem. 191, 295-298 (1951).

60. R. H. Burris. The Study of Growth Substances in Plant Metabolism. In, "Plant Growth Substances," pp. 93-95. F. Skoog, ed. The University of Wisconsin Press, Madison, 1951.

61. S. Grisolia, R. H. Burris and P. P. Cohen. Carbon Dioxide and Ammonia Fixation in the Biosynthesis of Citrulline. J. Biol. Chem. 191, 203-209 (1951).

62. I. H. Miller, Jr. and R. H. Burris. Effect of Plant Growth Substances upon Oxidation of Ascorbic and Glycolic Acids by Cell-free Enzymes from Barley. Amer. J. Bot. 38, 547-549 (1951).

63. A. C. Wagenknecht, A. J. Riker, T. C. Allen and R. H. Burris. Plant Growth Substances and the Activity of Cell-free Respiratory Enzymes. Amer. J. Bot. 38, 550-554 (1951).

64. F. S. Eberts, Jr., R. H. Burris and A. J. Riker. The Effects of Indole-3-acetic Acid and Common Organic Acids on the Respiration of Slices from Tomato Stem and Crown Gall Tissue. Amer. J. Bot. 38, 618-621 (1951).

65. I. Zelitch, E. D. Rosenblum, R. H. Burris and P. W. Wilson. Comparison of the Metabolism of Ammonia and Molecular Nitrogen in Clostridium. J. Bacteriol. 62, 747-752 (1951).

66. I. Zelitch, P. W. Wilson and R. H. Burris. The Amino Acid Composition and Distribution of N^{15} in Soybean Root Nodules Supplied N^{15}-enriched N_2. Plant Physiol. 27, 1-8 (1952).

67. V. Zbinovsky and R. H. Burris. Metabolism of Infiltrated Organic Acids by Tobacco Leaves. Plant Physiol. 27, 240-250 (1952).

68. R. H. Burris. Photosynthesis. Research Reviews, April (1952) pp. 21-25.

69. R. H. Burris and P. W. Wilson. Effect of Haemoglobin and Other Nitrogenous Compounds on the Respiration of the Rhizobia. Biochem. J. 51, 90-96 (1952).

70. M. H. Aprison and R. H. Burris. Time Course of Fixation of N_2 by Excised Soybean Nodules. Science 115, 264-265 (1952).
71. A. E. Williams and R. H. Burris. Nitrogen Fixation by Blue-green Algae and Their Nitrogenous Composition. Amer. J. Bot. 39, 340-342 (1952).
72. J. S. Wall, A. C. Wagenknecht, J. W. Newton and R. H. Burris. Comparison of the Metabolism of Ammonia and Molecular Nitrogen in Photosynthesizing Bacteria. J. Bacteriol. 63, 563-573 (1952).
73. L. A. Hyndman, R. H. Burris and P. W. Wilson. Properties of Hydrogenase from Azotobacter vinelandii. J. Bacteriol. 65, 522-531 (1953).
74. R. H. Burris. Organic Acids in Plant Metabolism. Ann. Rev. Plant Physiol. 4, 91-114 (1953). (Ann. Rev., Inc., Stanford, CA).
75. R. H. Burris. Carbon-14 Studies on Organic Acid Metabolism in Plants. Proceedings of the Fourth Annual Oak Ridge Summer Symposium, Aug. 25-30, 1952; p. 342-377. Available from Office of Technical Services, Dept. Commerce, Washington 25, D.C.
76. J. W. Newton, P. W. Wilson and R. H. Burris. Direct Demonstration of Ammonia as an Intermediate in Nitrogen Fixation by Azotobacter. J. Biol. Chem. 204, 445-451 (1953).
77. R. H. Burris. Studies of Biological Nitrogen Fixation with N^{15}. The Use of Isotopes in Plant and Animal Research. U.S. Atomic Energy Commission, pp. 68-80 in TID 5098, U.S. Govt. Printing Office, Washington, D.C. (1953).
78. D. O. Brummond and R. H. Burris. Transfer of C^{14} by Lupine Mitochondria through Reactions of the Tricarboxylic Acid Cycle. Proc. Nat. Acad. Sci. U.S.A. 39, 754-759 (1953).
79. P. W. Wilson and R. H. Burris. Biological Nitrogen Fixation--a Reappraisal. Ann. Rev. Microbiol. 7, 415-432 (1953).
80. V. Zbinovsky and R. H. Burris. New Techniques for Adding Organic Acids to Silicic Acid Columns. Anal. Chem. 26, 208-210 (1954).
81. R. W. Newburgh and R. H. Burris. Effect of Inhibitors on the Photosynthetic Fixation of Carbon Dioxide. Arch. Biochem. Biophys. 49, 98-109 (1954).
82. M. H. Aprison, W. E. Magee and R. H. Burris. Nitrogen Fixation by Excised Soybean Root Nodules. J. Biol. Chem. 208, 29-39 (1954).
83. F. S. Eberts, Jr., R. H. Burris and A. J. Riker. The Metabolism of Nitrogenous Compounds by Sunflower Crown Gall Tissue Cultures. Plant Physiol. 29, 1-10 (1954).
84. W. E. Magee and R. H. Burris. Fixation of N_2^{15} by Excised Nodules. Plant Physiol. 29, 199-200 (1954).
85. C. R. Noll, Jr. and R. H. Burris. Nature and Distribution of Glycolic Acid Oxidase in Plants. Plant Physiol. 29, 261-265 (1954).
86. D. G. Wilson, K. W. King and R. H. Burris. Transamination Reactions in Plants. J. Biol. Chem. 208, 863-874 (1954).
87. S. Grisolia and R. H. Burris. Preparation of Glutamate and Carbamyl Glutamate Selectively Labeled with Deuterium. J. Biol. Chem. 210, 109-117 (1954).
88. S. Grisolia, R. H. Burris and P. P. Cohen. Fate of Deutero-labeled Carbamyl Glutamate in Citrulline Biosynthesis. J. Biol. Chem. 210, 761-764 (1954).
89. R. R. Smeby, V. Zbinovsky, R. H. Burris and F. M. Strong. The Organic Acids of Narcissus poeticus. J. Amer. Chem. Soc. 76, 6127-6130 (1954).
90. M. M. Mozen and R. H. Burris. The Incorporation of ^{15}N-labeled Nitrous Oxide by Nitrogen Fixing Agents. Biochim. Biophys. Acta 14, 577-578 (1954).
91. W. E. Magee and R. H. Burris. Fixation of N_2 and Utilization of Combined Nitrogen by Nostoc muscorum. Amer. J. Bot. 41, 777-782 (1954).

92. D. O. Brummond and R. H. Burris. Reactions of the Tricarboxylic Acid Cycle in Green Leaves. J. Biol. Chem. 209, 755-765 (1954).
93. A. B. Krall and R. H. Burris. Evidence for the Participation of Cytochrome Oxidase in Photosynthetic Fixation of Carbon Dioxide. Physiologia Plantarum 7, 768-776 (1954).
94. A. I. Virtanen, T. Moisio, R. M. Allison and R. H. Burris. Fixation of Molecular Nitrogen by Excised Nodules of the Alder. Acta Chem. Scand. 8, 1730-1731 (1954).
95. R. H. Burris, W. E. Magee and M. K. Bach. The pN_2 and the pO_2 Function for Nitrogen Fixation by Excised Soybean Nodules. Ann. Acad. Sci. Fennicae, Ser. A, II. Chem. 60. A. I. Virtanen homage vol. 190-199 (1955).
96. J. S. Cohen and R. H. Burris. A Method for the Culture of Hydrogen Bacteria. J. Bacteriol. 69, 316-319 (1955).
97. M. M. Mozen and R. H. Burris. Experiments with Nitramide as a Possible Intermediate in Biological Nitrogen Fixation. J. Bacteriol. 70, 127-128 (1955).
98. A. I. Virtanen, T. Moisio and R. H. Burris. Fixation of Nitrogen by Nodules Excised from Illuminated and Darkened Pea Plants. Acta Chem. Scand. 9, 184-186 (1955).
99. M. M. Mozen, R. H. Burris, S. Lundbom and A. I. Virtanen. The Effect of Nitrous Oxide on Nitrate Utilization by Azotobacter vinelandii. Acta Chem. Scand. 9, 1232-1234 (1955).
100. R. H. Burris. Studies on the Mechanism of Biological Nitrogen Fixation. In, "Inorganic Nitrogen Metabolism," pp. 316-343. William D. McElroy and Bentley Glass, eds. The Johns Hopkins Press, Baltimore, 1956.
101. R. W. Scott, R. H. Burris and A. J. Riker. Nonvolatile Organic Acids of Crown Galls, Crown Gall Tissue Cultures and Normal Stem Tissue. Plant Physiol. 30, 355-360 (1955).
102. A. Tissieres and R. H. Burris. Purification and Properties of Cytochromes C_4 and C_5 from Azotobacter vinelandii. Biochem. Biophys. Acta 20, 436-437 (1956).
103. W. E. Magee and R. H. Burris. Oxidative Activity and Nitrogen Fixation in Cell-free Preparations from Azotobacter. J. Bacteriol. 71, 635-643 (1956).
104. R. M. Allison and R. H. Burris. Kinetics of Fixation of Nitrogen by Azotobacter vinelandii. J. Biol. Chem. 224, 351-364 (1957).
105. R. H. Burris. Nitrogen Fixation. A Conference on Radioactive Isotopes in Agriculture, U. S. Govt. Printing Office, Washington, D.C.; TID 7512, pp. 361-369, 1956.
106. G. E. Hoch, H. N. Little and R. H. Burris. Hydrogen Evolution from Soybean Root Nodules. Nature 179, 430-431 (1957).
107. D. P. Burma and R. H. Burris. Kinetics of Ammonia Utilization by Azotobacter vinelandii. J. Biol. Chem. 225, 287-295 (1957).
108. D. P. Burma and R. H. Burris. Metabolism of Nitrogen by Cell-free Preparations from Azotobacter vinelandii. J. Biol. Chem. 225, 723-733 (1957).
109. W. H. Peterson, R. H. Burris, R. Sant and H. N. Little. Production of Toxic Gas (Nitrogen Oxides) in Silage Making. Agr. Food Chem. 6, 121-126 (1958).
110. F. H. Bergmann, J. C. Towne and R. H. Burris. Assimilation of Carbon Dioxide by Hydrogen Bacteria. J. Biol. Chem. 230, 13-24, (1958).
111. M. K. Bach, W. E. Magee and R. H. Burris. Translocation of Photosynthetic Products to Soybean Nodules and Their Role in Nitrogen Fixation. Plant Physiol. 33, 118-124 (1958).
112. G. J. Fritz, W. G. Miller, R. H. Burris and L. Anderson. Direct Incorporation of Molecular Oxygen into Organic Material by Respiring Corn Seedlings. Plant Physiol. 33, 159-161 (1958).

113. N. Raggio, M. Raggio and R. H. Burris. Enhancement by Inositol of the Nodulation of Isolated Bean Roots. Science 129, 211-212 (1959).
114. J. B. Mudd, B. G. Johnson, R. H. Burris and K. P. Buchholtz. Oxidation of Indoleacetic Acid by Quackgrass Rhizomes. Plant Physiol. 34, 144-148 (1959).
115. N. Raggio, M. Raggio and R. H. Burris. Nitrogen Fixation by Nodules Formed on Isolated Bean Roots. Biochim. Biophys. Acta 32, 274-275 (1959).
116. R. H. Burris. Nitrogen Nutrition. Ann. Rev. Plant Physiol. 10, 301-328 (1959).
117. J. B. Mudd and R. H. Burris. Participation of Metals in Peroxidase-catalyzed Oxidations. J. Biol. Chem. 234, 2774-2777 (1959).
118. G. Delhumeau-Arrecillas and R. H. Burris. Effects of Azaserine on Azotobacter agilis. J. Bact. 78, 740-741 (1959).
119. T. Tamaoki, A. C. Hildebrandt, A. J. Riker, R. H. Burris and B. Hagihara. Oxidative and Phosphorylative Activities of Cytoplasmic Particles from Plant Tissue Cultures. Nature 184, 1491-1492 (1959).
120. J. B. Mudd and R. H. Burris. Inhibition of Peroxidase-catalyzed Oxidations. J. Biol. Chem. 234, 3281-3285 (1959).
121. N. P. Neumann and R. H. Burris. Cytochromes C_4 and C_5 of Azotobacter vinelandii: Chromatographic Purification, Crystallization, and a Study of Their Physical Properties. J. Biol. Chem. 234, 3286-3290 (1959).
122. G. E. Hoch, K. C. Schneider and R. H. Burris. Hydrogen Evolution and Exchange and Conversion of N_2O to N_2 by Soybean Root Nodules. Biochim. Biophys. Acta 37, 273-279 (1960).
123. R. N. Kurtzman, Jr., A. C. Hildebrandt, R. H. Burris and A. J. Riker. Inhibition and Stimulation of Tobacco Mosaic Virus by Purines. Virology 10, 432-448 (1960).
124. L. C. Wang and R. H. Burris. Mass Spectrometric Study of Nitrogenous Gases Produced by Silage. Ag. and Food Chem. 8, 239-242 (1960).
125. K. C. Schneider, C. Bradbeer, R. N. Singh, L. C. Wang, P. W. Wilson and R. H. Burris. Nitrogen Fixation by Cell-Free Preparations from Microorganisms. Proc. Nat. Acad. Sci. 46, 726-733 (1960).
126. T. Tamaoki, A. C. Hildebrandt, R. H. Burris, A. J. Riker and B. Hagihara. Respiration and Phosphorylation of Mitochondria from Normal and Crown-gall Tissue Cultures of Tomato. Plant Physiol. 35, 942-947 (1960).
127. R. H. Burris. Hydroperoxidases (Peroxidases and Catalases). In, "Handbuch der Pflanzenphysiologie-Encyclopedia of Plant Physiology," vol. 12, pp. 365-400. W. Ruhland, ed. 1960.
128. T. Tamaoki, A. C. Hildebrandt, R. H. Burris and A. J. Riker. Oxidation of Reduced Diphosphopyridine Nucleotide by Mitochondria from Normal and Crown-gall Tissue Cultures of Tomato. Plant Physiol. 36, 347-351 (1961).
129. L. C. Wang, J. Garcia-Rivera and R. H. Burris. Metabolism of Nitrate by Cattle. Biochem. J. 81, 237-242 (1961).
130. F. J. Bergersen, R. H. Burris and P. W. Wilson. Biochemical Studies on Soybean Nodules. Recent Advances in Botany, pp. 589-593, (1961).
131. J. E. McNary and R. H. Burris. Energy Requirements for Nitrogen Fixation by Cell-Free Preparations from Clostridium pasteurianum. J. Bacteriol. 84, 598-599 (1962).
132. A. R. Wasserman, J. C. Garver and R. H. Burris. Purification of Cytochrome C and Other Hemoproteins from Wheat Germ. Phytochemistry 2, 7-14 (1963).
133. R. H. Burris. Discussion of Evolution of Biological Nitrogen Fixation. In, "Proc. 5th Int. Cong. Biochemistry Vol. III." A. Oparin, ed. Pergamon Press Ltd., Oxford, England, 1963, pp. 173-177.
134. D. Wang and R. H. Burris. Carbon Metabolism of C^{14}-labeled Amino Acids in Wheat Leaves. II. Serine and Its Role in Glycine Metabolism. Plant Physiol. 38, 430-439 (1963).

135. R. D. Dua and R. H. Burris. Stability of Nitrogen-fixing Enzymes and the Reactivation of a Cold Labile Enzyme. Proc. Nat. Acad. Sci. 50, 169-175 (1963).
136. I. R. Hamilton, R. H. Burris and P. W. Wilson. Hydrogenase and Nitrogenase in a Nitrogen-fixing Bacterium. Proc. Nat. Acad. Sci. 52, 637-641 (1964).
137. R. H. Burris. Photosynthesis-reduction in the Presence of a Strong Oxidant. Vol. V. Molecular Structure and Biochem. Reactions; Proc. Robt. A. Welch Found. Conf. on Chem. Res. Dec. 4-6, 1961, pp. 113-116, published 1964.
138. I. R. Hamilton, R. H. Burris, P. W. Wilson and C. H. Wang. Pyruvate Metabolism, Carbon Dioxide Assimilation, and Nitrogen Fixation by an Archomobacter Species. J. Bacteriol. 89, 647-653 (1965).
139. D. Wang and R. H. Burris. Carbon Metabolism of C^{14}-labeled Amino Acids in Wheat Leaves. III. Further Studies on the Role of Serine in Glycine Metabolism. Plant Physiol. 40, 415-418 (1965).
140. D. Wang and R. H. Burris. Carbon Metabolism of Glycine and Serine in Relation to the Synthesis of Organic Acids and a Guanine Derivative. Plant Physiol. 40, 419-424 (1965).
141. A. R. Wasserman and R. H. Burris. Hemoprotein from Wheat Germ. Phytochemistry 4, 413-423 (1965).
142. I. R. Hamilton, R. H. Burris and P. W. Wilson. Pyruvate Metabolism by a Nitrogen-fixing Bacterium. Biochem. J. 96, 383-389 (1965).
143. T. O. Munson, M. J. Dilworth and R. H. Burris. Method for Demonstrating Cofactor Requirements for Nitrogen Fixation. Biochim. Biophys. Acta 104, 278-281 (1965).
144. M. J. Dilworth, D. Subramanian, T. O. Munson and R. H. Burris. The Adenosine Triphosphate Requirement for Nitrogen Fixation in Cell-free Extracts of Clostridium pasteurianum. Biochim. Biophys. Acta 99, 486-503 (1965).
145. R. D. Dua and R. H. Burris. Studies of Cold Lability and Purification of a Nitrogen-activating Enzyme. Biochim. Biophys. Acta 99, 504-510 (1965).
146. P. Plengvidhya and R. H. Burris. Inhibitors of Photophosphorylation and Photoreduction. Plant Physiol. 40, 997-1002 (1965).
147. R. H. Burris, H. C. Winter, T. O. Munson and J. Garcia-Rivera. Intermediates and Cofactors in Nitrogen Fixation. In, "Non-heme Iron Protein: Role in Energy Conversion." A symposium sponsored by the Chas. F. Kettering Research Laboratory, The Antioch Press, Yellow Springs, Ohio, 1965, pp. 315-321.
147a. R. H. Burris. Chairman's Remarks. Non-heme Iron Proteins: Role in Energy Conversion. A. San Pietro, ed. Antioch Press, Yellow Springs, Ohio, 1965, pp. 241-242.
148. A. Lockshin and R. H. Burris. Inhibitors of Nitrogen Fixation in Extracts from Clostridium pasteurianum. Biochim. Biophys. Acta 111, 1-10 (1965).
149. R. H. Burris. Biological Nitrogen Fixation. Ann. Rev. Plant Physiol. 17, 155-184 (1966).
150. R. H. Burris. Nitrogen Fixation. In, "Plant Biochemistry," pp. 961-979. J. Bonner and J. E. Varner, eds. Academic Press, New York, 1965.
151. A. Lockshin and R. H. Burris. Solubilization and Properties of Chloroplast Lamellar Protein. Proc. Natl. Acad. Sci. 56, 1564-1570 (1966).
152. S. M. C. Dietrich and R. H. Burris. Effect of Exogenous Substrates on the Endogenous Respiration of Bacteria. J. Bacteriol. 93, 1467-1470 (1967).
153. R. V. Klucas and R. H. Burris. Locus of Nitrogen Fixation in Soybean Nodules. Fixation by Crushed Nodules. Biochim. Biophys. Acta 136, 399-401 (1967).

154. J. Garcia-Rivera and R. H. Burris. Hydrazine and Hyroxylamine as Possible Intermediates in the Biological Fixation of Nitrogen. Arch. Biochem. Biophys. 119, 167-172 (1967).
155. R. Schöllhorn and R. H. Burris. Reduction of Azide by the N_2-fixing Enzyme System. Proc. Natl. Acad. Sci. 57, 1317-1323 (1967).
156. R. Schöllhorn and R. H. Burris. Acetylene as a Competitive Inhibitor of N_2 Fixation. Proc. Natl. Acad. Sci. 58, 213-216 (1967).
157. M. Kelley, R. V. Klucas and R. H. Burris. Fractionation and Storage of Nitrogenase from Azotobacter vinelandii. Biochem. J. 105, 3c-5c (1967).
158. W. D. P. Stewart, G. P. Fitzgerald, and R. H. Burris. In situ Studies on N_2 Fixation using the Acetylene Reduction Technique. Proc. Natl. Acad. Sci. 58, 2071-2078 (1967).
159. H. C. Winter and R. H. Burris. Stoichiometry of the Adenosine Triphosphate Requirement for N_2 Fixation and H_2 Evolution by a Partially Purified Preparation of Clostridium pasteurianum. J. Biol. Chem. 243, 940-944 (1968).
160. F. S. Fang and R. H. Burris. Cytochrome c in Hydrogenomonas eutropha. J. Bacteriol. 96, 298-305 (1968).
161. W. D. P. Stewart, G. P. Fitzgerald and R. H. Burris. Acetylene Reduction by Nitrogen-fixing Blue-green Algae. Arch. für Mikrobiol. 62, 336-348 (1968).
162. T. O. Munson and R. H. Burris. Nitrogen Fixation by Rhodospirillum rubrum Grown in Nitrogen-limited Continuous Culture. J. Bacteriol. 97, 1093-1098 (1969).
163. R. H. Burris. Progress in the Biochemistry of Nitrogen Fixation. Proc. Roy. Soc. B. 172, 339-354 (1969).
164. R. T. Swank and R. H. Burris. Restoration by Ubiquinone of Azotobacter vinelandii Reduced Nicotinamide Adenine Dinucleotide Oxidase Activity. J. Bacteriol. 98, 311-313 (1969).
165. D. E. Dravnieks, F. Skoog and R. H. Burris. Cytokinin Activation of de novo Thiamine Biosynthesis in Tobacco Callus Cultures. Plant Physiology 44, 866-870 (1969).
166. R. T. Swank and R. H. Burris. Purification and Properties of Cytochromes c of Azotobacter vinelandii. Biochim. Biophys. Acta 180, 473-489 (1969).
167. C. A. Ouelette, R. H. Burris and P. W. Wilson. Deoxyribonucleic Acid Base Composition of Species of Klebsiella, Azotobacter and Bacillus. Antonie van Leeuwenhoek, J. Microbiol. and Serol. 35, 275-286 (1969).
168. J.-P. Vandecasteele and R. H. Burris. Purification and Properties of the Constituents of the Nitrogenase Complex from Clostridium pasteurianum. J. Bacteriol. 101, 794-801 (1970).
169. W. D. P. Stewart, G. P. Fitzgerald and R. H. Burris. Acetylene Reduction Assay for Determination of Phosphorus Availability in Wisconsin Lakes. Proc. Natl. Acad. Sci. 66, 1104-1111 (1970).
170. D. Kleiner and R. H. Burris. The Hydrogenase of Clostridium pasteurianum. Kinetic Studies and the Role of Molybdenum. Biochim. Biophys. Acta 212, 417-427 (1970).
171. J. E. Sundquist and R. H. Burris. Light-dependent Structural Changes in the Lamellar Membranes of Isolated Spinach Chloroplasts: Measurement by Electron Microscopy. Biochim. Biophys. Acta 223, 115-121 (1970).
172. D. Rusness and R. H. Burris. Acetylene Reduction (Nitrogen Fixation) in Wisconsin Lakes. Limnology and Oceanography 15, 808-813 (1970).
173. Y. I. Shethna, N. A. Stombaugh and R. H. Burris. Ferredoxin from Bacillus polymyxa. Biochem. Biophys. Res. Commun. 42, 1108-1116 (1971).
174. W. D. P. Stewart, T. Mague, G. P. Fitzgerald and R. H. Burris. Nitrogenase Activity in Wisconsin Lakes of Differing Degrees of Eutrophication. New Phytol. 70, 497-509 (1971).

175. R. H. Burris. Fixation by Free-living Micro-organisms: Enzymology. In, "The Chemistry and Biochemistry of Nitrogen Fixation," pp. 105-160. J. R. Postgate, ed. Plenum Press, London, 1971.
176. T. Ljones and R. H. Burris. A Continuous Spectrophotometric Assay for Nitrogenase. Anal. Biochem. 45, 448-452 (1972).
177. T. H. Mague and R. H. Burris. Reduction of Acetylene and Nitrogen by Field-grown Soybeans. New Phytol. 71, 275-281 (1972).
178. R. H. Burris. Nitrogen Fixation - Assay Methods and Techniques. In, "Methods in Enzymology 24B." A. San Pietro, ed.; S. P. Colowick and N. O. Kaplan, eds.-in-chief. Academic Press, New York and London, 1972.
179. P. P. Wong and R. H. Burris. Nature of Oxygen Inhibition of Nitrogenase from Azotobacter vinelandii. Proc. Natl. Acad. Sci. U.S.A. 69, 672-675 (1972).
180. J. F. Kratochvil, R. H. Burris, M. K. Seikel and J. M. Harkin. Isolation and Characterization of α-Guaiaconic Acid and the Nature of Guaiacum Blue. Phytochem. 10, 2529-2531 (1971).
181. M.-Y. W. Tso, T. Ljones and R. H. Burris. Purification of the Nitrogenase Proteins from Clostridium pasteurianum. Biochim. Biophys. Acta 267, 600-604 (1972).
182. T. Ljones and R. H. Burris. ATP Hydrolysis and Electron Transfer in the Nitrogenase Reaction with Different Combinations of the Iron Protein and the Molybdenum-iron Protein. Biochim. Biophys. Acta 275, 93-101 (1972).
183. W. H. Orme-Johnson, W. D. Hamilton, T. Ljones, M.-Y. W. Tso, R. H. Burris, V. K. Shah and W. J. Brill. Electron Paramagnetic Resonance of Nitrogenase and Nitrogenase Components from Clostridium pasteurianum W5 and Azotobacter vinelandii OP. Proc. Natl. Acad. Sci. U.S.A. 69, 3142-3145 (1972).
184. J. C. Hwang and R. H. Burris. Nitrogenase-catalyzed Reactions. Biochim. Biophys. Acta 283, 339-350 (1972).
185. J. C. Hwang, C. H. Chen and R. H. Burris. Inhibition of Nitrogenase-catalyzed Reductions. Biochim. Biophys. Acta 292, 256-270 (1973).
186. L. N. Vanderhoef, B. Dana, D. Emerich and R. H. Burris. Acetylene Reduction in Relation to Levels of Phosphate and Fixed Nitrogen in Green Bay. New Phytol. 71, 1097-1105 (1972).
187. T. H. Mague and R. H. Burris. Biological Nitrogen Fixation in the Great Lakes. BioScience 23, 236-239 (1973).
188. M.-Y. W. Tso and R. H. Burris. The Binding of ATP and ADP by Nitrogenase Components from Clostridium pasteurianum. Biochim. Biophys. Acta 309, 263-270 (1973).
189. N. A. Stombaugh, R. H. Burris and W. H. Orme-Johnson. Ferredoxins from Bacillus polymyxa. Low Potential Iron-sulfur Proteins which Appear to Contain Single Four Iron, Four Sulfur Centers Accepting a Single Electron on Reduction. J. Biol. Chem. 248, 7951-7956 (1973).
190. W. H. Campbell, W. H. Orme-Johnson and R. H. Burris. A Comparison of the Physical and Chemical Properties of Four Cytochromes c from Azotobacter vinelandii. Biochem. J. 135, 617-630 (1973).
191. R. H. Burris. Methodology. In, "The Biology of Nitrogen Fixation," pp. 9-33. A. Quispel, ed. North-Holland Publishing Company, Amsterdam-Oxford, 1974.
192. R. H. Burris and W. H. Orme-Johnson. Survey of Nitrogenase and Its EPR Properties. In, "Microbial Iron Metabolism, A Comprehensive Treatise," pp. 187-209. J. B. Neilands, ed. Academic Press, Inc., New York, 1974.
193. R. H. Burris. Biological Nitrogen Fixation, 1924-1974. Plant Physiol. 54, 443-449 (1974).
194. J. M. Rivera-Ortiz and R. H. Burris. Interactions among Substrates and Inhibitors of Nitrogenase. J. Bacteriol. 123, 537-545 (1975).

195. D. L. Erbes, R. H. Burris and W. H. Orme-Johnson. On the Iron-sulfur Cluster in Hydrogenase from Clostridium pasteurianum W5. Proc. Nat. Acad. Sci. U.S.A. 72, 4795-4799 (1975).
196. R. H. Burris. The Acetylene-reduction Techniques. In, "Nitrogen Fixation by Free-living Micro-organisms," pp. 249-257. W. D. P. Stewart, ed. IBP volume 6, Cambridge Univ. Press, Cambridge, 1975.
197. R. H. Burris. Preparation and Properties of Nitrogenase Proteins. In, "Nitrogen Fixation by Free-living Micro-organisms," pp. 333-349. W. D. P. Stewart, ed. IBP volume 6, Cambridge Univ. Press, Cambridge, 1975.
198. R. H. Burris. Nitrogen Fixation by Blue-green Algae of the Lizard Island Area of the Great Barrier Reef. Aust. J. Plant Physiol. 3, 41-51 (1976).
199. R. B. Peterson and R. H. Burris. Properties of Heterocysts Isolated with Colloidal Silica. Arch. Microbiol. 108, 35-40 (1976).
200. R. B. Peterson and R. H. Burris. Conversion of Acetylene Reduction Rates to Nitrogen Fixation Rates in Natural Populations of Blue-green Algae. Anal. Biochem. 73, 404-410 (1976).
201. H. C. Winter and R. H. Burris. Nitrogenase. Ann. Rev. Biochem. 45, 409-426 (1976).
202. Y. Okon, S. L. Albrecht and R. H. Burris. Factors Affecting Growth and Nitrogen Fixation of Spirillium lipoferum. J. Bacteriol. 127, 1248-1254 (1976).
203. R. H. Burris and W. H. Orme-Johnson. Mechanism of Biological N_2 Fixation. In, "Proc. of the 1st International Symposium on Nitrogen Fixation." W. E. Newton and C. J. Nyman, eds. Washington State Univ. Press, 1976, pp. 208-233.
204. N. A. Stombaugh, J. E. Sundquist, R. H. Burris and W. H. Orme-Johnson. Oxidation-reduction Properties of Several Low Potential Iron-sulfur Proteins and of Methylviologen. Biochemistry 15, 2633-2641 (1976).
205. R. H. Burris and C. C. Black, Editors. CO_2 Metabolism and Plant Productivity. University Park Press, Baltimore, MD, 1976, 431 pages.
206. P. W. Ludden and R. H. Burris. Activating Factor for the Iron Protein of Nitrogenase from Rhodospirillum rubrum. Science 194, 424-426 (1976).
207. Y. Okon, S. L. Albrecht and R. H. Burris. Carbon and Ammonia Metabolism of Spirillum lipoferum. J. Bacteriol. 128, 592-597 (1976).
208. J. D. Tjepkema and R. H. Burris. Nitrogenase Activity Associated with Some Wisconsin Prairie Grasses. Plant and Soil 45, 81-94 (1976).
209. R. H. Burris. Nitrogen Fixation. In, "Plant Biochemistry," pp. 887-908. J. Bonner and J. E. Varner, eds. Academic Press, New York, 1976.
210. Y. Okon, S. L. Albrecht and R. H. Burris. Methods for Growing Spirillum lipoferum and for Counting it in Pure Culture and in Association with Plants. Appl. Environ. Microbiol. 33, 85-88 (1977).
211. C. B. Osmond, M. M. Bender and R. H. Burris. Pathways of CO_2 Fixation in the CAM plant Kalachoë daigremontiana. III. Correlation with $\delta^{13}C$ Value during Growth and Water Stress. Aust. J. Plant Physiol. 3, 787-799 (1976).
212. D. W. Emerich and R. H. Burris. Interactions of Heterologous Nitrogenase Components that Generate Catalytically Inactive Complexes. Proc. Nat. Acad. Sci. U.S.A. 73, 4369-4373 (1976).
213. R. B. Peterson, E. E. Friberg and R. H. Burris. Diurnal Variation in N_2 Fixation and Photosynthesis by Aquatic Blue-green Algae. Plant Physiol. 59, 74-80 (1977).
214. Y. Okon, J. P. Houchins, S. L. Albrecht and R. H. Burris. Growth of Spirillum lipoferum at Constant Partial Pressures of Oxygen, and the Properties of its Nitrogenase in Cell-free Extracts. J. Gen. Microbiol. 98, 87-93 (1977).

215. R. H. Burris. Energetics of Biological N_2 Fixation. In, "Biological Solar Energy Conversion," pp. 275-289. A. San Pietro and S. Tamura, eds. Academic Press, New York, 1977.
216. R. H. Burris. Overview of Nitrogen Fixation. In, "Genetic Engineering for Nitrogen Fixation," pp. 9-18. A. Hollaender, R. H. Burris, P. R. Day, R. W. F. Hardy, D. R. Helinski, M. R. Lamborg, L. Owens and R. C. Valentine, eds. Plenum Press, New York and London, 1977.
217. R. H. Burris, Y. Okon and S. L. Albrecht. Physiological Studies of Spirillum lipoferum. In, "Genetic Engineering for Nitrogen Fixation," pp. 445-450. A. Hollaender, R. H. Burris, P. R. Day, R. W. F. Hardy, D. R. Helinski, M. R. Lamborg, L. Owens and R. C. Valentine, eds. Plenum Press, New York and London, 1977.
218. R. H. Burris. A Synthesis Paper on Nonleguminous N_2-fixing Systems. In, "Recent Developments in Nitrogen Fixation," pp. 487-511. W. Newton, J. R. Postgate and C. Rodriguez-Barrueco, eds. Academic Press, London, New York, San Francisco, 1977.
219. S. L. Albrecht, Y. Okon and R. H. Burris. Effects of Light and Temperature on the Association between Zea mays and Spirillum lipoferum. Plant Physiol. 60, 528-531 (1977).
220. T. Ljones and R. H. Burris. Evidence for Oneelectron Transfer by the Fe Protein of Nitrogenase. Biochem. Biophys. Res. Commun. 80, 22-25, (1978).
221. J. Döbereiner, R. H. Burris, A. Hollaender, A. A. Franco, C. A. Neyra, and D. B. Scott, eds. Limitations and Potentials for Biological Nitrogen Fixation in the Tropics, p. 398. Plenum Press, New York and London, 1978.
222. R. H. Burris, T. Ljones and D. W. Emerich. Nitrogenase Systems. In "Limitations and Potentials for Biological Nitrogen Fixation in the Tropics," pp. 191-207. (See above reference.)
223. R. H. Burris, S. L. Albrecht and Y. Okon. Physiology and Biochemistry of Spirillum lipoferum. In, "Limitations and Potentials for Biological Nitrogen Fixation in the Tropics," pp. 303-315. (See above reference.)
224. R. B. Peterson and R. H. Burris. Hydrogen Metabolism in Isolated Heterocysts of Anabaena 7120. Arch. Microbiol. 116, 125-132 (1978).
225. R. H. Burris. Overview of Biological N_2 Fixation. In, "Genetic Engineering for Nitrogen Fixation," pp. 21-30. Alexander Hollaender, ed. U.S. Government Printing Office, Washington, D.C., 20402; Stock No. 038-000-00354-2, 1978.
226. T. Ljones and R. H. Burris. Nitrogenase: The Reaction Between the Fe Protein and Bathophenanthrolinedisulfonate as a Probe for Interactions with MgATP. Biochemistry 17, 1866-1872 (1978).
227. W. Lockau, R. B. Peterson, C. P. Wolk and R. H. Burris. Modes of Reduction of Nitrogenase in Heterocysts Isolated from Anabaena Species. Biochim. Biophys. Acta 502, 298-308 (1978).
228. D. W. Emerich and R. H. Burris. Complementary Functioning of the Component Proteins of Nitrogenase from Several Bacteria. J. Bacteriol. 134, 936-943 (1978).
229. R. H. Burris. Advances in Biological Nitrogen Fixation. Developments in Industrial Microbiology 19, 1-13 (1978). [Proc. 34th General Meeting of the Society for Industrial Microbiology, L. A. Underkofler, ed., 624 pp.]
230. R. V. Hageman and R. H. Burris. Nitrogenase and Nitrogenase Reductase Associate and Dissociate with each Catalytic Cycle. Proc. Natl. Acad. Sci. U.S.A. 75, 2699-2702 (1978).
231. D. L. Erbes and R. H. Burris. The Kinetics of Methyl Viologen Oxidation and Reduction by the Hydrogenase from Clostridium pasteurianum. Biochim. Biophys. Acta 525, 45-54 (1978).
232. R. H. Burris and R. B. Peterson. Nitrogen-fixing Blue-green Algae: Their H_2 Metabolism and Their Activity in Freshwater Lakes. In,

"Environmental Role of Nitrogen-fixing Blue-green Algae and Asymbiotic Bacteria," pp. 28-40. U. Granhall, ed. Ecol. Bull. (Stockholm) 26, 1978.

233. R. H. Burris, Y. Okon and S. L. Albrecht. Properties and Reactions of Spirillum lipoferum., In "Environmental Role of Nitrogen-fixing Blue-green Algae and Asymbiotic Bacteria," pp. 353-363. U. Granhall, ed. Ecol. Bull. (Stockholm) 26, 1978.

234. D. W. Emerich and R. H. Burris. Nitrogenase from Bacillus polymyxa. Purification and Properties of the Component Proteins. Biochim. Biophys. Acta 536, 172-183 (1978).

235. D. W. Emerich and R. H. Burris. Preparation of Nitrogenase. In, "Methods in Enzymology," vol. 53, Biomembranes, Part D: Biological Oxidations; Mitochondrial and Microbial Systems, pp. 314-329. S. Fleischer and L. Packer, eds. Academic Press, New York, San Francisco and London, 1978.

236. R. V. Hageman and R. H. Burris. Kinetic Studies on Electron Transfer and Interaction between Nitrogenase Components from Azotobacter vinelandii. Biochemistry 17, 4117-4124 (1978).

236a. R. H. Burris. Future of Biological N_2 Fixation. BioScience 28, 563 (1978).

237. P. W. Ludden, Y. Okon and R. H. Burris. The Nitrogenase System of Spirillum lipoferum. Biochem. J. 173, 1001-1003 (1978).

238. P. W. Ludden and R. H. Burris. Purification and Properties of Nitrogenase from Rhodospirillum rubrum, and Evidence for Phosphate, Ribose and an Adenine-like Unit Covalently Bound to the Iron Protein. Biochem. J. 175, 251-259 (1978).

239. D. W. Emerich, T. Ljones and R. H. Burris. Nitrogenase: Properties of the Catalytically Inactive Complex Between the Azotobacter vinelandii MoFe Protein and the Clostridium pasteurianum Fe Protein. Biochim. Biophys. Acta 527, 359-369 (1978).

240. R. H. Burris. The Early Biochemistry. In, "A Treatise on Dinitrogen Fixation, Sections I and II: Inorganic and Physical Chemistry and Biochemistry, Chapter 1, " pp. 383-398. R. W. F. Hardy, F. Bottomeley and R. C. Burns, eds. John Wiley & Sons, Inc., New York, 1979.

241. R. H. Burris. Inhibition. In, "A Treatise on Dinitrogen Fixation, Sections I and II: Inorganic and Physical Chemistry and Biochemistry" Chapter 5, pp. 569-604. R. W. F. Hardy, F. Bottomeley and R. C. Burns, eds. John Wiley & Sons, Inc., New York, 1979.

242. D. K. Stumpf and R. H. Burris. A Micromethod for the Purification and Quantification of Organic Acids of the Tricarboxylic Acid Cycle in Plant Tissues. Anal. Biochem. 95, 311-315 (1979).

243. D. R. Benson, D. J. Arp, and R. H. Burris. Cell-free Nitrogenase and Hydrogenase from Actinorhizal Root Nodules. Science 205, 688-689 (1979).

244. D. J. Arp and R. H. Burris. Purification and Properties of the Particulate Hydrogenase from the Bacteroids of Soybean Root Nodules. Biochim. Biophys. Acta 570, 221-230 (1979).

245. P. W. Ludden and R. H. Burris. Removal of an Adenine-like Molecule during Activation of Dinitrogenase Reductase from Rhodospirillum rubrum. Proc. Natl. Acad. Sci. U.S.A. 76, 6201-6205 (1979).

246. R. V. Hageman and R. H. Burris. Changes in the EPR Signal of Dinitrogenase from Azotobacter vinelandii during the Lag Period before Hydrogen Evolution Begins. J. Biol. Chem. 254, 11189-11192 (1979).

247. L. C. Davis, M. T. Henzl, R. H. Burris and W. H. Orme-Johnson. Iron-sulfur Clusters in the Molybdenum-iron Protein Component of Nitrogenase. Electron Paramagnetic Resonance of the Carbon Monoxide Inhibited State. Biochemistry 18, 4860-4869 (1979).

248. R. B. Peterson and R. H. Burris. Nitrogen Fixation by Blue-green Algae in Lakes. In, "Advances in Cyanophyte Research," pp. 15-24. R. N. Singh Memorial Volume, S. P. Singh, D. N. Tiwari, A. K. Kashyap

and P. K. Yadava, eds. Dept. of Botany, Banaras Hindu Univ., Varsanasi-221005, India, 1978.

249. R. H. Burris and R. V. Hageman. Electron Partitioning from Dinitrogenase to Substrate and the Kinetics of ATP Utilization. In, "Molybdenum Chemistry of Biological Significance," pp. 23-37. W. E. Newton and S. Otsuka, eds. Plenum Press, New York, 1980.
250. R. V. Hageman and R. H. Burris. Nitrogenase: Electron Transfer and Allocation and the Role of ATP. In, "Molybdenum and Molybdenum-Containing Enzymes," pp. 403-426. M. P. Coughlan, ed. Pergamon Press, Oxford, 1980.
251. R. H. Burris. The Global Nitrogen Budget--Science or Seance? In, "Nitrogen Fixation," 1, pp. 7-16. W. E. Newton and W. H. Orme-Johnson, eds. University Park Press, Baltimore, 1980.
252. D. R. Benson, D. J. Arp, and R. H. Burris. Hydrogenase in Actinorhizal Root Nodules and Root Nodule Homogenates. J. Bacteriol. 142, 138-144 (1980).
253. M. H. Spalding, D. K. Stumpf, M. S. B. Ku, R. H. Burris and G. E. Edwards. Crassulacean Acid Metabolism and Diurnal Variations in Internal CO_2 and O_2 Concentrations in Sedum praealtum DC. Aust. J. Plant Physiol. 6, 557-567 (1979).
254. R. V. Hageman and R. H. Burris. Electron Allocation to Alternative Substrates of Azotobacter Nitrogenase is Controlled by the Electron Flux through Dinitrogenase. Biochim. Biophys. Acta 591, 63-75 (1980).
255. R. V. Hageman. W. H. Orme-Johnson and R. H. Burris. Role of Magnesium Adenosine 5'-Triphosphate in the Hydrogen Evolution Reaction Catalyzed by Nitrogenase from Azotobacter vinelandii. Biochemistry 19, 2333-2342 (1980).
256. R. V. Hageman and R. H. Burris. Electrochemistry of Nitrogenase and the Role of ATP. In, "Current Topics in Bioenergetics, 10," pp. 279-291. D. R. Sanadi, ed. Academic Press, New York, 1980.
257. R. H. Burris, D. J. Arp., D. R. Benson, D. W. Emerich, R. V. Hageman, T. Ljones, P. W. Ludden, and W. J. Sweet. The Biochemistry of Nitrogenase. In, "Nitrogen Fixation," pp. 37-54. W. D. P. Stewart and J. R. Gallon, eds. Academic Press, London, 1980.
258. W. J. Sweet and R. H. Burris. Inhibition of Nitrogenase Activity by NH_4^+ in Rhodospirillum rubrum. J. Bacteriol. 145, 824-831 (1981).
259. D. W. Emerich, R. V. Hageman and R. H. Burris. Interactions of Dinitrogenase and Dinitrogenase Reductase. In, "Advances in Enzymology and Related Areas of Molecular Biology," pp. 1-22. A. Meister, ed. John Wiley and Sons, New York, 1981.
260. J. P. Houchins and R. H. Burris. Occurrence and Localization of Two Distinct Hydrogenases in the Heterocystous Cyanobacterium Anabaena sp. Strain 7120. J. Bacteriol. 146, 209-214 (1981).
261. J. P. Houchins and R. H. Burris. Comparative Characterization of Two Distinct Hydrogenases from Anabaena sp. Strain 7120. J. Bacteriol. 146, 215-221 (1981).
262. D. J. Arp and R. H. Burris. Kinetic Mechanism of the Hydrogen-oxidizing Hydrogenase from Soybean Nodule Bacteroids. Biochemistry 20, 2234-2240 (1981).
263. S. L. Albrecht, Y. Okon, J. Lonnquist and R. H. Burris. Nitrogen Fixation by Corn-Azospirillum Associations in a Temperate Climate. Crop Science 21, 301-306 (1981).
264. A. Hochman and R. H. Burris. Effect of Oxygen on Acetylene Reduction by Photosynthetic Bacteria. J. Bacteriol. 147, 492-499 (1981).
265. R. H. Burris, D. J. Arp, R. V. Hageman, J. P. Houchins, W. J. Sweet and M.-Y. W. Tso. Mechanism of Nitrogenase Action. In, "Current Perspectives in Nitrogen Fixation," pp. 56-66. A. H. Gibson and W. E. Newton, eds. Australian Academy of Science, Canberra, 1981.
266. J. P. Houchins and R. H. Burris. Light and Dark Reactions of the Uptake Hydrogenase in Anabaena 7120. Plant Physiol. 68, 712-716 (1981).

267. J. P. Houchins and R. H. Burris. Physiological Reactions of the Reversible Hydrogenase from Anabaena 7120. Plant Physiol. 68, 717-721 (1981).
268. D. K. Stumpf and R. H. Burris. Organic Acid Contents of Soybean: Age and Source of Nitrogen. Plant Physiol. 68, 989-991 (1981).
269. D. K. Stumpf and R. H. Burris. Biosynthesis of Malonate in Roots of Soybean Seedlings. Plant Physiol. 68, 992-995 (1981).
270. P. W. Ludden and R. H. Burris. In vivo and in vitro Studies on ATP and Electron Donors to Nitrogenase in Rhodospirillum rubrum. Arch. Microbiol. 130, 155-158 (1981).
271. P. W. Ludden, R. V. Hageman, W. H. Orme-Johnson and R. H. Burris. Properties and Activities of "Inactive" Iron Protein from Rhodospirillum rubrum. Biochim. Biophys. Acta 700, 213-216 (1982).
272. D. J. Arp and R. H. Burris. Isotope Exchange and Discrimination by the Hydrogen-oxidizing Hydrogenase from Soybean Root Nodules. Biochim. Biophys. Acta 700, 7-15 (1982).
273. W. J. Sweet and R. H. Burris. Effects of in vivo Treatments on the Activity of Nitrogenase Isolated from Rhodospirillum rubrum. Biochim. Biophys. Acta 680, 17-21 (1982).
274. L. S. Privalle and R. H. Burris. In situ Effects of ADP/ATP Levels on Nitrogenase Activity in Isolated Heterocysts of Anabaena 7120. In, "Energy Coupling in Photosynthesis." Proc. 11th Steenbock Symp., July 6-8, 1981, Madison, WI. B. R. Selman and S. Selman-Reimer, eds. Elsevier, North-Holland, 1981, pp. 353-358.
275. R. H. Burris. History and Projections for Biological N_2 Fixation. In, "Current Topics in Plant Biochemistry and Physiology," pp. 180-194. D. D. Randall, D. G. Blevins and R. Larson, eds. University of Missouri-Columbia, 1983.
276. L. S. Privalle and R. H. Burris. Adenine Nucleotide Levels in and Nitrogen Fixation by the Cyanobacterium Anabaena sp. Strain 7120. J. Bacteriol. 154, 351-355 (1983).
277. J.-L. Li and R. H. Burris. Influence of pN_2 and pD_2 on HD Formation by Various Nitrogenases. Biochemistry 22, 4472-4480 (1983).
278. J. H. Guth and R. H. Burris. The Role of Mg^{2+} and Mn^{2+} in the Enzyme-catalyzed Activation of Nitrogenase Fe Protein from Rhodospirillum rubrum. Biochem. J. 213, 741-749 (1983).
279. L. S. Privalle and R. H. Burris. Permeabilization of Isolated Heterocysts of Anabaena sp. Strain 7120 with Detergent. J. Bacteriol. 155, 940-942 (1983).
280. R. H. Burris. Uptake and Assimilation of $^{15}N_4{}^{+}$ by A Variety of Corals. Marine Biology 75, 151-155 (1983).
281. J. H. Guth and R. H. Burris. Comparative Study of the Active and Inactive Forms of Dinitrogenase Reductase from Rhodospirillum rubrum. Biochim. Biophys. Acta 749, 91-100 (1983).
282. J. H. Guth and R. H. Burris. Inhibition of Nitrogenase-catalyzed NH_3 Formation by H_2. Biochemistry 22, 5111-5122 (1983).
283. R. H. Burris. Enzymology of Nitrogenase, p. 143-144 and J. H. Guth and R. H. Burris. Inhibition of N_2 Fixation by H_2, p. 148. In, "Advances in Nitrogen Fixation Research." C. Veeger and W. E. Newton, eds. Martinus Nijhoff/Dr W. Junk Publishers, the Hague, 1984.
284. C. van Kessel and R. H. Burris. Effect of H_2 Evolution on $^{15}N_2$ Fixation, C_2H_2 Reduction and Relative Efficiency of Leguminous Symbionts. Physiol. Plant. 59, 329-334 (1983).
285. L. S. Privalle and R. H. Burris. D-Erythrose Supports Nitrogenase Activity in Isolated Anabaena sp. Strain 7120 Heterocysts. J. Bacteriol. 157, 350-356 (1984).
286. Burris, R. H. Nitrogen Metabolism in the Coral-Algal Symbiosis. Proc. Am. Philosophical Soc. 128, 85-92 (1984).
287. Simpson, F. B. and R. H. Burris. A Nitrogen Pressure of 50 Atmospheres Does Not Prevent Evolution of Hydrogen by Nitrogenase. Science 224, 1095-1097 (1984).

288. Martinez-Drets, G., M. Del Gallo, C. Burpee and R. H. Burris. Catabolism of Carbohydrates and Organic Acids and Expression of Nitrogenase by Azospirilla. J. Bacteriol. 159, 80-85 (1984).

AUTHOR INDEX

SUBJECT INDEX